Numerical Groundwater Modeling

Flow and Contaminant Migration

WILLIAM C. WALTON

LEWIS PUBLISHERS

national water well association

Library of Congress Cataloging-in-Publication Data

Walton, William Clarence.
Numerical groundwater modeling: flow and contaminant migration
p. cm.
Bibliography: p.
Includes index.
1. Hydrogeology—Computer programs. 2. GWFL3D (Computer program) 3. GWTR3D (Computer program) I. Title.
GB1001.72.E45W35 1989 551.49'072'4—dc19 88-23008

ISBN 0-87371-196-3

Second Printing 1989

LEWIS PUBLISHERS, INC.
121 South Main Street, Chelsea, Michigan 48118

PRINTED IN THE UNITED STATES OF AMERICA

Preface

Two microcomputer programs for simulating quasi-three-dimensional groundwater flow and contaminant migration are presented in this book. The numerical and analytical techniques incorporated in these programs are described. Target readers are practicing engineers, geologists, and hydrogeologists; water well contractors; educators and students; governmental personnel; industry representatives; and others. The reader is expected to have a working knowledge of hydrogeology, have microcomputer operating experience, and have read books on groundwater hydrology and modeling such as those written by Marsily (1986) and Bear and Verruijt (1987).

The two programs GWFL3D and GWTR3D are extensive modifications and extensions of, and sequels to, the popular public domain mainframe computer programs PLASM and RANDOM WALK presented in the following two publications: Prickett and Lonnquist (1971) and Prickett et al. (1981). These two publications are required reading because they provide program information not repeated herein. GWFL3D and GWTR3D are tailored to the IBM PC (trademark of International Business Machines Corp.) and compatible microcomputers. They are prompt/option driven, contain no menus, and feature user-friendly interactive data base entry and frequent screen displays of helpful data base information. Aquifer system conditions are user-defined by entering an appropriate number designated for desired options in response to prompt message lines. Input/output options include: data base hard copies produced by a printer and data base diskette files; spacial and temporal numerical water level and concentration diskette files, screen displays, and hard copies; and XY and XYZ diskette output files formatted for export to commercial graphics programs.

I hope this publication will assist practitioners in solving groundwater flow and contamination problems by integrating simulation techniques scattered, until now, throughout the literature and in several programs in the public domain. The modular and simple structure of GWFL3D and GWTR3D, although not especially efficient, should enable the reader to further modify and extend the programs as desired with little difficulty. Data entry has been simplified so that the user can run the programs without worrying about FORTRAN input file structures and editors. The production of graphic displays has been facilitated and data base calculations external to the program have been eliminated. It should be emphasized that the programs contained herein are useful only within the context of the assumptions and simplifications on which they are based.

This book grew out of my frequent use and admiration for the simplicity of PLASM and RANDOM WALK, and my periodic need to modify those two programs to fit consulting assignments. My long-time professional association with Tom Prickett and Phil Davis, especially over the last 10 years, inspired me to tackle computer programming and to complete this publication. My wife, Ellen, taught me to use the word processor effectively, and my poor drafting ability forced me to learn how to draw with the aid of a microcomputer (see Voisinet, 1986).

William C. Walton

Mahomet, Illinois

William C. Walton is a semiretired Consultant in Water Resources with a home on the banks of the Sangamon River at Mahomet, Illinois. Bill received his BS in Civil Engineering from Lawrence Institute of Technology in 1948 and attended Indiana University, the University of Wisconsin, Ohio State University, and Boise State University. He served for 10 years as Director of the Water Resources Research Center and Professor of Geology and Geophysics at the University of Minnesota.

Bill's 40 years of experience in the water resources field include 1 year with the U. S. Bureau of Reclamation, 7 years with the U.S. Geological Survey, 6 years with the Illinois State Water Survey, and 9 years with consulting companies such as Camp Dresser & McKee Inc. and Geraghty & Miller. He was Executive Director of the Upper Mississippi River Basin Commission for 5 years and has participated in water resources projects throughout the United States and Canada, and in Haiti, El Salvador, Libya, and Saudi Arabia.

The positions he has held include: Vice President of the National Water Well Association; editor of the journal

Ground Water; Chairman of the Ground Water Committee of the Hydraulics Division, American Society of Civil Engineers; member of the U.S. Geological Survey Advisory Committee on Water Data for Public Use; Consultant to the Office of Science and Technology, Washington, D.C.; and Advisor to the United States Delegation to the Coordinating Council of the International Hydrological Decade of UNESCO.

Bill also served as a Visiting Scientist for the American Geophysical Union and the American Geological Institute and has lectured at many universities throughout the United States. He is the author of 5 books and over 70 technical papers.

Contents

List of Figures

List of Tables

1

Introduction

The two microprograms GWFL3D and GWTR3D introduced in this book simulate multi-aquifer production wells, flowing wells, partially penetrating production wells, well storage capacity, aquifer layering, aquitard storativity, delayed gravity yield under water table conditions, drains, mines, areal contaminant leakage sinks, and areal contaminant leakage sources, in addition to the aquifer system and facilities simulated and described in PLASM and RANDOM WALK. Extension techniques and modifications and revised program operation of PLASM and RANDOM WALK included in GWFL3D and GWTR3D are described in detail herein. The strengths and weaknesses of PLASM and RANDOM WALK are discussed in detail by Marsily (1986, pp. 339–340) and Kinzelbach (1987, pp. 656–674), respectively.

Similar to PLASM, GWFL3D solves large sets of finite-difference equations for one-, two-, and quasi-three-dimensional nonsteady flow in heterogeneous aquifer systems with a modified alternating direction implicit algorithm. The program provides numerical spacial-water level and time-water level output options, and simulates the following aquifer system and facility features: constant or variable production well discharge or injection well recharge; water table, leaky artesian, and nonleaky artesian conditions; constant head, constant flux, and no-flow aquifer system boundaries; induced surface water infiltration; groundwater evapotranspiration; aquifer storativity conversion; and multi-aquifer conditions. In simulating induced surface water infiltration, GWFL3D allows the user to change the stream width, length, and surface elevation, and the streambed hydraulic conductivity and thickness, each time increment. It is assumed that there is always enough water in the stream to supply the aquifer and that the liquid's density and viscosity remain constant in space and time, as in PLASM.

Similar to RANDOM WALK, GWTR3D simulates one, two, and quasi-three-dimensional contaminant transport in groundwater, and employs discrete parcel random walk techniques. Contaminant transport is based on a particle in a cell technique for advective mechanisms, and a random walk technique for dispersion mechanisms. The program simulates continuous and slug contaminant source areas of various shapes, contaminant sinks such as wells and streams, vertically averaged salt-water fronts, and contaminant leakage from overlying source beds. It is assumed that contaminant concentration does not affect the liquid's density and viscosity. Processes of contaminant sorption and decay are simulated. Dissimilar to RANDOM WALK, GWTR3D decouples the flow model from the solute transport model, calls for a head data base, and simulates one time increment per program run.

Emphasis has been placed in this book on practice rather than theory. Simulation equations are presented in abbreviated and final form. Mathematical derivations of equations may be pursued through references. Subject matter is presented in as nonmathematical a manner as possible.

GWFL3D and GWTR3D were verified with public domain programs presented by McDonald and Harbaugh (1984) and Konikow and Bredehoeft (1978), and data and equations presented by Prickett and Lonnquist (1971, pp. 12–14, 17–19, 27–28, 30–31, 34–35, 37–38, 40–41, 43–44, 46–47); Prickett et al. (1981, pp.71–90); Trescott et al. (1976, p. 7); Bennett et al. (1982, pp. 338–339); Rushton and Redshaw (1979, pp. 246–248, 263–266); and Bear (1972, pp. 366–367).

Several commercial programs were available in 1988 for compiling GWFL3D and GWTR3D including: Microsoft FORTRAN v. 4.0 (trademark of Microsoft Corp.), Ryan-McFarland FORTRAN v. 2.4 (trademark of PC Brand), Lahey FORTRAN v. 2.2 (trademark of Lahey Computer Systems, Inc.), and IBM Professional FORTRAN v. 1.22 (trademark of International Business Machines Corp.). These compilers implement FORTRAN 77 (standard ANSI X309–1978), support Assembly and Pascal or C language routine calls, support 8087 and 80287 coprocessors, facilitate porting of mini- and mainframe computers, and offer IBM, VS, DEC VAX (trademark of Digital Equipment Corp.) extensions. GWFL3D and GWTR3D were written specifically for Microsoft FORTRAN v. 3.2; with slight modification they will work with other compilers.

The three diskettes included with this book contain the source and executable files for GWFL3D and GWTR3D. Source codes are listed in Appendixes A and B. Complete instructions on how to run the programs are presented in Appendix C. The diskettes run on microcomputers such as the IBM PC (trademark of International Business Machines Corp.) and compatible machines that use the MS-DOS (trademark of Microsoft Corp.) operating system. Model capabilities and limitations are summarized in Appendix D to prevent potential program misuse. Example input/output displays are presented in Appendix E to guide the user through program operation.

Commercial software packages were available in 1988 for producing graphs, contour maps, and 3-D plots from output data generated by GWFL3D and GWTR3D. For example, GRAPHER (trademark of Golden Software, Inc., P.O. Box 281, Golden, CO 80402) displays black and white or color linear-linear, log-linear, linear-log, and log-log XY graphs. It supports the IBM PC, XT, or AT (or compatibles) and a wide variety of printers and plotters. GRAPHER has five different types of curve fitting to control points: linear, logarithmic, exponential, power, and cubic spline. Grid lines may be solid or dashed and multiple scales may be displayed on a single graph. Axes may be positioned at any location and axis tick marks and labels are user-defined. Lines through data points may be solid or dashed. Other features include: user-defined data point symbols, data file (free ASCII) import or keyboard spreadsheet data entry, data point labels, superposition of graphs, zoom and panning, text selected from user-defined symbols sets, and text rotated to any angle and scaled to any size.

SURFER (trademark of Golden Software, Inc.) creates contour maps and 3-D surfaces. It imports ASCII data files or creates a spreadsheet data base from the keyboard. Inverse distance or kriging techniques are available for interpolation between irregularly spaced data points. SURFER displays black and white or color contour maps with: in-line contour labels scaled to any size and user-defined symbol sets; bold, dashed, or normal contour lines; hachure marks; smoothed or unsmoothed contour lines; irregular contour intervals; user-defined scaling and sizing; location posted by user-defined symbol sets and labeled with numeric or text strings; axes with tick marks and labels on any side of the map; and multiple shaped boundaries. Black and white or color 3-D surface plots may be created with: stacked contours or fishnet surfaces; perspective or orthographic projections; hidden line removal; rotation to any angle; smoothed or unsmoothed lines; irregular stacked contour intervals; user-defined scaling and sizing; locations posted by any symbol and labeled with numeric or text strings; 3-D axes with tick marks and labels; multiple shaped boundaries; and the top, bottom, or both of the surface. A surface plot from a user-defined function may be displayed. SURFER supports IBM PC, XT, or AT (or compatibles) and a wide variety of printers and plotters.

GRAPHER and SURFER have an “import” option to load (X,Y,Z) or (X,Y) data from ASCII (text) files generated under other programs or created with a word processor or an editor such as IBM’s EDLIN editor on the PC-DOS (trademark of International Business Machines Corp.) diskette. Files with a .DAT filename extension must contain one

(X,Y,Z) or (X,Y) entry per line; values on a single line must be separated by either a space, tab, or comma. GWFL3D and GWTR3D contain options to create files with formats matching those required by GRAPHER and SURFER.

PLOT 88 (trademark of Plotworks, Inc., Dept. W-9, 16440 Eagles Crest Road, Ramona, CA 92065) contains a library of graphic subroutines which may be called for by programs written in Microsoft, Ryan-McFarland, Lahey F77L, or IBM Professional FORTRAN when constructing XY graphs, contour maps, and 3-D mesh drawings. PLOT 88 supports IBM PC, XT, or AT (or compatibles) and a wide range of printers and plotters. It contains fully integrated sets of subroutines for generating contour maps and 3-D mesh surfaces. Subroutines draw: a linear or logarithmic axis with title, tick marks, and tick labels; an arithmetic, log-log, or semilog line between a set of XY points; a smooth curve between a set of XY points; characters and special centered symbols; two-dimensional surface contours based on a grid of control points; and a three-dimensional mesh surface based on a grid of control points. Other subroutines generate straight line segments; define graph enlargement or reduction, line width, width-to-height ratio of characters, format of axis tick, color index, shading pattern, and line type and symbol characteristics; select a character set; determine logarithmic scale factors; perform two-dimensional interpolation to construct a grid of data points; smooth gridded data; determine contour levels; and mark and label randomly spaced data points.

The reader is referred to the International Ground Water Modeling Center (IGWMC), Holcomb Research Institute, Butler University for current listings of other available microcomputer graphics and groundwater programs. Some program information is also provided in the National Water Well Association (NWWA) Publications Catalog.

In addition to commercial graphics software packages, microcomputer graphics programs are listed in several publications. For example, a linear XY graph IBM PC BASIC program is listed by Hearn and Baker (1983, p. 62). Linear, semilog, and log-log XY graph APPLE II BASIC (trademark of Apple Computer, Inc.) programs are listed by Korites (1982, pp. Lin/Lin 6–14, Log/Lin 3–12, and Log/Log 2–8). IBM PC BASIC versions of these programs are available on diskette. A 3-D surface plot IBM PC BASIC program is listed by Fowler (1984, pp. 292–296). Contours subroutines, written in IBM PC BASIC and QuickBASIC (trademark of Microsoft, Inc.), are listed by Simons (1983, pp. 487–492) and Bourke (1987, pp. 143–150). The theory of curve and surface fitting is described by Lancaster and Salkauskas (1986) and techniques for computer contouring of irregularly spaced data are described by Crain (1970); Pelto et al. (1968); and McLain (1974).

2

Groundwater Flow Program Operation

In GWFL3D, variable arrays are dimensioned, then program information and variable definitions are presented through subroutine INFO1. Supported English units are: gpd, cfs, days, min, ft, ft^2, ft^3, gpd/ft, and gpd/ft^2. Unit requirements are cited in data base input statements. Some useful unit conversions are: 1 L/sec = 15.85 US gal/min, 1 m/sec = 2.119 x 10^6 US gal/day ft^2, 1 m^2/sec = 6.954 x 10^6 US gal/day ft, 1 cm/sec = 2.121 x 10^4 US gal/day ft^2, 1 ft/day = 7.481 US gal/day ft^2, and 1 ft = 0.3048 m. The user may choose interactive keyboard time-constant data base specification or importation of a time-constant data base file through subroutine FILEI. With keyboard time-constant data base specification, the user defines time increments and the finite-difference grid (number of columns and rows and default grid spacing). Variable grid spacings within selected grid blocks may be user-defined.

The user may specify simulation of nonleaky, leaky, or water table conditions in the uppermost aquifer or aquifer layer. If appropriate, the constant head in the sourcebed above the uppermost aquifer or aquifer layer is user-defined.

In subroutine INFO2, the user may choose to simulate aquifer layering, aquitard storativity, and/or delayed gravity yield under water table conditions. The user specifies the number of aquifer units or aquifer layers based on information displayed on the screen concerning aquifer layering, aquitard storativity, delayed gravity yield, and aquifer system boundaries.

The user fills the following variable arrays with default values to simulate homogeneous conditions: aquifer or aquifer layer horizontal hydraulic conductivity, vertical hydraulic conductivity, storativity, and specific yield; aquifer thickness and top and base elevations; and aquitard vertical hydraulic conductivity and thickness. Information is displayed on the typical range of these variables; selected variables are then internally initialized. The user may revise variable values at nodes within selected grid blocks to simulate heterogeneous conditions and aquifer system boundaries.

The user specifies the number and coordinates of production wells; number and coordinates of injection wells; number and coordinates of flowing wells; flowing well radius, and top of casing elevation (through subroutine FWEL); and number and coordinates of time-drawdown observation wells. The flowing well simulation factor is calculated in subroutine FWEL. The nodal head convergence error tolerance and water balance error tolerance are user-specified. Aquifer, aquifer layer, and aquitard hydraulic characteristic values are adjusted for any variable grid spacings in the MAIN program. The user may choose to obtain a hard copy of grid spacings produced by a printer, store the time-constant data base in a file on a diskette through subroutine FILEO, and obtain a hard copy of the time-constant data base through subroutine DATPR. If the time-constant data base was imported, the keyboard time-constant data base entry would be bypassed and the program would continue at this point.

The time increment loop is started and the time-variable portion of the data base is interactively entered from the keyboard by the user. The production well discharge rate, injection well recharge rate, and uniform nodal recharge rate are user-defined. Subroutine MINE displays information concerning mine simulation and the user may specify the number of mine nodes, mine node coordinates, and the mine floor elevation. The user may choose active mine nodes, inactive mine nodes, backfilled mine nodes, and fully or partially penetrating mine nodes.

Storage factors at active mine nodes are set at 1E21 to create constant heads at mine nodes, and the heads at mine nodes are set at user-defined mine floor elevations. Active mine node transmissivities are adjusted for the mine cavity depending upon the depth of mine penetration which is user-defined. If the time increment is identified by the user as being the time increment during which the mine node first became active, the mine cavity dewatering rate may be calculated, storage factors at any inactive mine nodes are reduced to normal values, and mine node transmissivity adjustments are relaxed. The hydraulic characteristics of the backfill materials are user-defined at any backfilled mine nodes and transmissivity adjustments at these nodes are relaxed.

The number and coordinates of any streambed nodes, streambed width, length, thickness, and streambed vertical hydraulic conductivity; stream surface elevation; and streambed base elevation are user-defined in subroutine RIVER. Streambed nodal simulation factors are calculated (see Prickett and Lonnquist, 1971, pp. 33–36) and streambed nodes are identified.

The number and coordinates of any drain nodes, drain base elevation, drain backfill thickness, drain backfill effective hydraulic conductivity, drain diameter, and drain length are user-defined in subroutine DRAIN. Drain nodal simulation factors are calculated and drain nodes are identified.

Potential evapotranspiration and elevation below land surface at which evapotranspiration ceases are user-defined in subroutine EVAP (see Prickett and Lonnquist, 1971, pp. 33–36).

If aquitard storativity is simulated, the elapsed time after the last production well discharge rate change is user-defined. Aquitard storativity nodal simulation factors are calculated. The user may choose to cap flowing wells, in which case the flowing well simulation factor at the flowing well node is reset at zero. If necessary, the simulation factor is then set to a normal value.

The G and B arrays are initialized and heads for the next time increment are predicted (see Prickett and Lonnquist, 1971, pp. 11–48) in subroutine SIMEQ. The effective storativity and recharge rate simulating delayed gravity yield are calculated if necessary. At this point in the program, the user may choose to simulate a multi-aquifer production well. The production well coordinates, uppermost and lowermost aquifer units or layers open to the well, well radius, and well discharge rate are user-defined.

The convergence error and iteration counter are initialized and the flow rate term expressions associated with any multi-aquifer production well are generated. The finite-difference simultaneous equations are solved in subroutine SIMEQ with the iterative alternating direction implicit method (see Prickett and Lonnquist, 1971, pp. 3–7,46–48). Storativity conversion and transmissivity decreases, under water table conditions, are simulated in flow rate equations (see Prickett and Lonnquist, 1971, pp. 40–45).

Water balance components are calculated and displayed on the screen upon request of the user in subroutine WATBL, called from subroutine SIMEQ. The head error is compared with the head error tolerance, and iterations are continued until head convergence is acceptable (see Prickett and Lonnquist, 1971, pp. 14–15). Heads at any time-drawdown observation wells are calculated. Heads in production wells, discharges from flowing wells and mines, water balance components, composite water level or drawdown in any multi-aquifer production well, and multi-aquifer production well discharges from aquifer units or aquifer layers are calculated in subroutine RESLT. The user may choose to obtain a hard copy of these calculation results.

Calculated values of head or drawdown at nodes at the end of each time increment and in each aquifer or aquifer layer may be displayed on the screen in a plan view X (column number), Y (row number), Z (head or drawdown) format. The user may choose to create a hard copy of the screen display and simulate partially penetrating wells in aquifers, but not aquifer layers, in subroutine PARPEN. Well storage capacity may be simulated in subroutine WELSTOR. An X (X-coordinate of node), Y (Y-coordinate of node), Z (Z-coordinate of node) file on a diskette with a format specified in commercial graphics programs may be created through subroutine XYZDT. A hard copy of profile view heads or drawdowns in an X (node number), Y (aquifer unit or aquifer layer number), Z (head or drawdown) format may be obtained through subroutine PROFIL. The user may choose a screen display of time-head or drawdown tables for any selected observation wells. A hard copy of the screen display may be created. The user may choose to create a file on a diskette with an X (time), Y (head or drawdown) format specified in commercial graphics programs.

3

Groundwater Flow Simulation Techniques

Techniques in GWFL3D algorithms for simulating leakage through aquitards, induced streambed infiltration, evapotranspiration, storativity conversion, water table conditions, and multiple aquifer systems are those in PLASM as described by Prickett and Lonnquist (1971, pp. 30–52). The details of these techniques will not be repeated herein. Simulation techniques in GWFL3D not covered in PLASM are described below, together with brief reviews of selected fundamental simulation information presented in PLASM.

Time Increments and Grid Spacing

Time increments and grid spacings are interdependent and should be jointly specified so that head changes at nodes adjacent to sink/source nodes are appreciable during the initial time increment. In GWFL3D, the number of time increments must be less than 51 unless delayed gravity yield is simulated, in which case the number of time increments must be less than 21 because of limited microcomputer core storage. Nonuniform time increments are user-defined as part of the time-variable data base. The proper choice of time increments is important for high precision head or drawdown calculations. Small time increments should be specified whenever there is a change in sink/source or aquifer system boundary conditions, such as a production well starting or stopping. Time increments should then be gradually increased as a geometric progression of ratio 1.2 until the next change when small time increments are required again (Prickett and Lonnquist, 1971, pp. 15–17).

Time increments should be small in comparison to the length of the simulation period. To ensure high precision head or drawdown calculations, six time increments without significant changes in sink/source and aquifer system boundary conditions should precede the time of interest (see Marsily, 1986, pp. 398–400; Prickett and Lonnquist, 1971, pp. 13–14). In general, the smaller the initial time increment, the greater the precision of head or drawdown calculations.

The finite-difference grid lines superposed over the aquifer system map are indexed using the I (column), J (row) notation colinear with the X and Y directions, respectively. Coordinate axes intersect at the upper-left corner of the map. I coordinates increase left to right and J coordinates increase top to bottom. Grid spacing is centered at the nodes (grid line intersections). In multi-aquifer or aquifer layer systems, units or layers are numbered from land surface downward using the K notation colinear with the Z direction. The origin of the system is the upper-left corner of the topmost unit or layer. Grid spacing is the same in all aquifer system units or layers. In GWFL3D, the number of aquifer units and/or aquifer layers must be less than 5 because of limited microcomputer core storage (640 K).

Leaky artesian conditions with a constant head in the sourcebed may be simulated in the uppermost unit or layer by specifying the sourcebed head and aquitard hydraulic conductivity and thickness. Nonleaky artesian or water table conditions may be simulated in the uppermost unit or layer by setting aquitard hydraulic conductivity and thickness to zero, thus creating an aquiclude. By default, aquitard storativity is assumed to be negligible. If aquifer layering is simulated, it is assumed that water table condi-

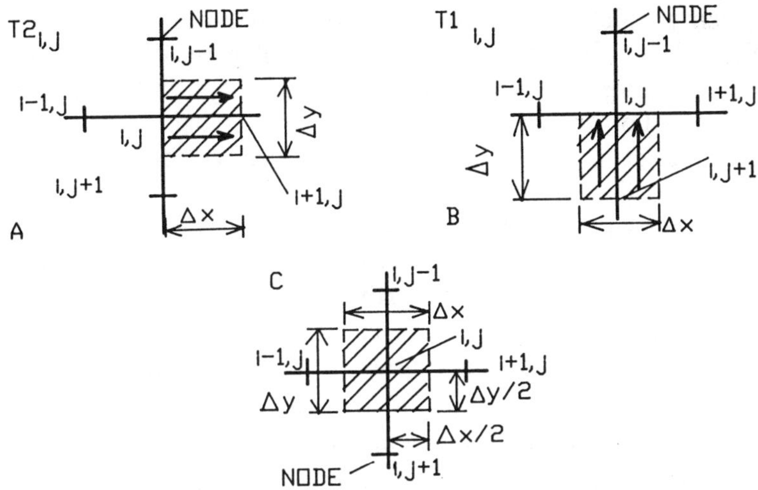

Figure 1. Vector volumes for aquifer or aquifer layer transmissivities (A and B) and storage and simulation factors (C).

tions prevail in the uppermost layer at the top of the aquifer system. If delayed gravity yield is simulated, the number of units or layers must be 1.

Horizontal projections of vector volumes for transmissivity, storativity, and source and sink simulation factors (Q_n) are defined in Figure 1. These vector volumes, except that related to Q_n, extend the full depth of the aquifer or aquifer layer.

The precision of head or drawdown calculations is related to the grid spacing. The grid spacing should be small in comparison to the simulation area. Smallest grid spacings should be specified where the head gradient varies significantly such as around production or injection wells, rivers, mines, and drains. Grid spacing should be gradually increased outward from these features. In GWFL3D, the number of columns or rows is limited to 20 because of limited microcomputer core storage, making it desirable to specify variable grid spacing in many cases with large grid spacings along grid borders. Grid spacings are uniform along individual columns and rows. The ratio of adjacent grid spacings is usually less than 5, except for high precision head or drawdown calculations, when it is 2 or less (see Trescott et al., 1976, p. 30). Most nodes will be active with proper grid spacing.

Similar to PLASM, the algorithm in GWFL3D for solving a large set of finite-difference equations is based on a constant (square grid). The finite-difference equation for a model aquifer system with the square grid illustrated in Figure 2 is as follows (Prickett and Lonnquist, 1971, p. 4):

$$\begin{aligned} Q_n &+ T_{i-1,j,2}(h_{i-1,j} - h_{i,j}) + T_{i,j,1}(h_{i,j+1} - h_{i,j}) \\ &= T_{i,j,2}(h_{i,j} - h_{i+1,j}) + T_{i,j-1,1}(h_{i,j} - h_{i,j-1}) \\ &\quad + S\,\Delta X_i\,\Delta Y_j(h_{i,j} - ho_{i,j})/\Delta t + Q_{i,j} \end{aligned} \tag{1}$$

where

$T_{i,j,1}$ = aquifer or aquifer layer transmissivity within vector volume between nodes i,j and i,j+1

$T_{i,j,2}$ = aquifer or aquifer layer transmissivity within vector volume between nodes i,j and i+1,j

$h_{i,j}$ = calculated head at end of simulation time increment at node i,j

$ho_{i,j}$ = calculated head at end of previous simulation time increment at node i,j

Δt = simulation time increment

S = aquifer or aquifer layer storativity

$Q_{i,j}$ = constant withdrawal rate at node i,j

Q_n = sink or source flow rate at node i,j

ΔX_i = grid spacing in X-direction

ΔY_j = grid spacing in Y-direction

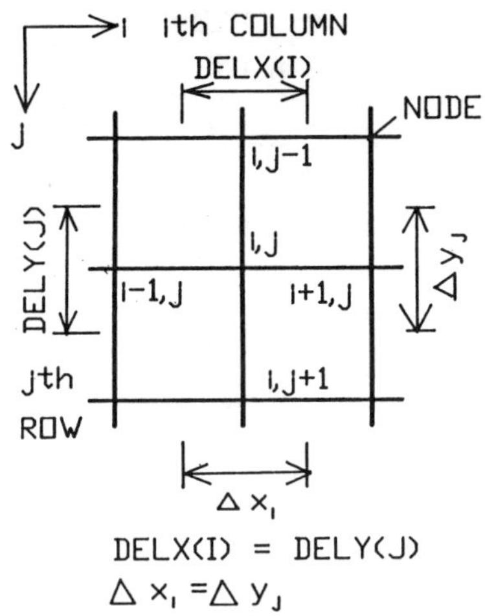

Figure 2. Uniform finite-difference grid.

In contrast, the finite-difference equation for a model aquifer system with the variable grid illustrated in Figure 3 is as follows [OSM (Office of Surface Mining), 1981, p. II–26]:

$$
\begin{aligned}
&Q_n + T_{i-1,j,2}(h_{i-1,j} - h_{i,j})[2\Delta Y_j/(\Delta X_{i+1} + \Delta X_i)] \\
&\quad + T_{i,j,1}(h_{i,j+1} - h_{i,j})[2\Delta X_i/(\Delta Y_j + \Delta Y_{j+1})] \\
&\quad = T_{i,j,2}(h_{i,j} - h_{i+1,j})[2\Delta Y_j/(\Delta X_i + \Delta X_{i+1})] \\
&\quad + T_{i,j-1,1}(h_{i,j} - h_{i,j-1})[2\Delta X_i/(\Delta Y_j + \Delta Y_{j+1})] \\
&\qquad + S\,\Delta X_i\,\Delta Y_j(h_{i,j} - ho_{i,j})/\Delta t + Q_{i,j} \qquad (2)
\end{aligned}
$$

Comparison of Equations 1 and 2 indicates that the algorithm for solving large sets of finite-difference equations based on a square grid can be utilized with variable grids, provided $T_{i,j,2}$ and $T_{i-1,j,2}$ are multiplied by the arithmetic mean expression $2\,\Delta Y_i/(\Delta X_{i+1} + \Delta X_i)$ and $T_{i,j,1}$ and $T_{i,j-1,1}$ are multiplied by $2\,\Delta X_i/(\Delta Y_{j+1} + \Delta Y_j)$. Similar to PLASM,

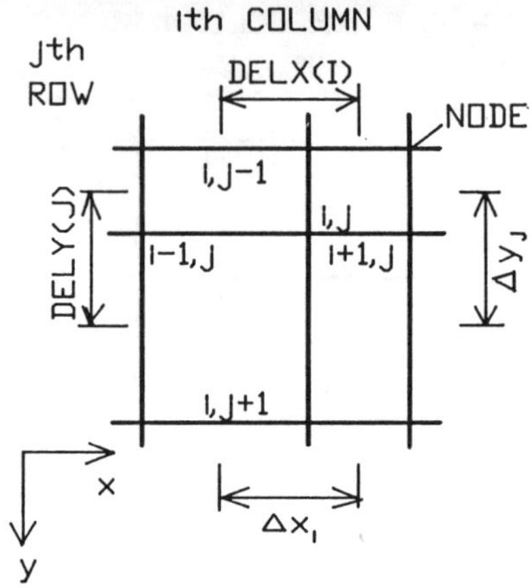

Figure 3. Variable finite-difference grid.

GWFL3D simulates variable grids by adjusting transmissivities accordingly.

Irregular Barrier and Recharge Boundaries

Barrier aquifer system (no flow) boundaries may be simulated by assigning zero transmissivities to nodes along boundaries. Recharge (constant head) boundaries may be simulated by assigning large (1E20, do not use 1E21) storage factors to nodes along boundaries (see Franke et al., 1985, pp. 1–26). Without transmissivity, storage factor, and simulation factor modifications along grid borders, aquifer system boundaries are located midway between adjacent nodes outside and inside boundaries. This location may be changed and irregular boundaries may be simulated with "vector volume" techniques described by Karplus (1958). The following equations are used to adjust transmissivities, storage factors, and simulation factors:

$$T_{xa} = T_{xu}[(\Delta Y/2 + Y_b)/(\Delta X/2 + X_b)] \quad (3)$$

$$T_{ya} = T_{yu}[(\Delta X/2 + X_b)/(\Delta Y/2 + Y_b)] \quad (4)$$

$$SF1_a = SF1_u[(\Delta X/2 + X_b)(\Delta Y/2 + Y_b)/(\Delta X\ \Delta Y)] \quad (5)$$

$$R_a = R_u[(\Delta X/2 + X_b)(\Delta Y/2 + Y_b)/(\Delta X\ \Delta Y)] \quad (6)$$

where

T_{xa} = aquifer or aquifer layer transmissivity in X-direction adjusted for irregular boundary location in gpd/ft

T_{ya} = aquifer or aquifer layer transmissivity in Y-direction adjusted for irregular boundary location in gpd/ft

T_{xu} = aquifer or aquifer layer transmissivity without boundary in X-direction in gpd/ft

T_{yu} = aquifer or aquifer layer transmissivity without boundary in Y-direction in gpd/ft

$SF1_a$ = aquifer or aquifer layer storage factor adjusted for irregular boundary adjacent to node in gal/ft

$SF1_u$ = aquifer or aquifer layer storage factor without boundary in gal/ft

R_a = simulation factor adjusted for irregular boundary adjacent to node in gpd/ft

R_u = simulation factor without boundary in gal/ft

ΔX = grid spacing in X-direction in ft

ΔY = grid spacing in Y-direction in ft

X_b = distance between grid line and boundary in X-direction in ft

Y_b = distance between grid line and boundary in Y-direction in ft

If there is no boundary in the Y-direction, $Y_b = \Delta Y/2$. If there is no boundary in the X-direction, $X_b = \Delta X/2$. With a square grid $\Delta X = \Delta Y$.

Water Balance

Recent versions of PLASM use the change in head between iterations summed for all nodes and the water balance summed for all nodes (see OSM, 1981, p. II–28) as measures of convergence error. Iterations are terminated when the convergence errors do not exceed specified error tolerances. If water balance is used as a measure of convergence, the precision of water balances at all nodes must be high. However, precision tends to be low at nodes where large simulation factors or storativities are multiplied by very small head losses or differences (Prickett, 1987, personal communication). This can cause severe program problems. For this reason, GWFL3D does not use water balance to measure convergence, but instead only displays water balances when requested to do so by the user.

Production Well Head

Similar to PLASM, GWFL3D simulates a production well in subroutine SIMEQ by imposing a discharge rate on a grid block centered at a node. Grid block dimensions are usually much larger than the production well diameter. The head numerically calculated at the production well node is an average head for the grid block, not the head in the production well (see Beljin, 1987, pp. 340–351). The concept of equivalent well block radius with a square (uniform) or rectangular (variable) grid and anisotropic hydraulic conductivity (Peaceman, 1983, pp. 531–543) based on the Thiem equation, may be used to calculate the head in the production well through subroutine RESLT.

First, the radius at which the head in the aquifer is equal to the numerically calculated head for the production well grid block is estimated with the following equation (Peaceman, 1983):

$$r_e = 0.28[(T_{yy}/T_{xx})^{1/2}\,\Delta X^2 + (T_{xx}/T_{yy})^{1/2}\,\Delta Y^2]^{1/2} \,/[(T_{yy}/T_{xx})^{1/4} + (T_{xx}/T_{yy})^{1/4}] \qquad (7)$$

where

r_e = equivalent well block radius in ft

T_{xx} = aquifer or aquifer layer transmissivity in X-direction at well node in gpd/ft

T_{yy} = aquifer or aquifer layer transmissivity in Y-direction at well node in gpd/ft

ΔX = well block grid spacing in X-direction in ft

ΔY = well block grid spacing in Y-direction in ft

Then the head in the production well based on the head numerically calculated for the production well block is estimated with the following equation (Peaceman, 1983):

$$h_w = h_{gb} - Q\ln(r_e/r_w)/[2\pi(T_{xx}T_{yy})^{1/2}] \qquad (8)$$

where

h_w = head in production well in ft

h_{gb} = average head in well block in ft

Q = production well discharge rate in gpd

r_w = production well effective radius in ft

Commonly, the grid spacing at the production well node is square and usually the ratio of grid spacings in the X- and Y-directions is less than 5 at the production well node. A method developed by Pedrosa and Aziz (1986, pp. 611–621) may be used to estimate the head in the production well with greater precision than can be attained with Equations 7 and 8.

The production well head may be adjusted for the effects of turbulent flow (well loss) with the following equation (Jacob, 1947, pp. 1047–1070):

$$s_{w1} = C_{w1}Q^2 \tag{9}$$

where

s_{w1} = additional drawdown due to well loss in ft

C_{w1} = well loss constant in sec^2/ft^5

Q = production well discharge rate in ft^3/sec

Variables in Equations 7–9 are user-defined in subroutine RESLT.

Partially Penetrating Wells

Similar to PLASM, GWFL3D in algorithms within subroutine SIMEQ assumes wells fully penetrate the aquifer or aquifer layer. Whether or not partially penetrating wells should be simulated may be determined with the following equation (Hantush, 1964, p. 566):

$$r_{pp} = 1.5m(P_H/P_V)^{1/2} \tag{10}$$

where

r_{pp} = distance beyond which partial penetration impacts are negligible in ft

m = aquifer thickness in ft

P_H = aquifer horizontal hydraulic conductivity in gpd/sq ft

P_V = aquifer vertical hydraulic conductivity in gpd/sq ft

Partially penetrating wells in aquifers may be simulated in GWFL3D through subroutine PARPEN with the following equation (see Hantush, 1964, pp. 347–358):

$$s_t = s + s_p \tag{11}$$

where

$$s_p = 114.6QW[u,r(P_V/P_H)^{1/2}/m,l/m,d/m,lo/m,do/m] \tag{12}$$

$$u = 1.87r^2S/(Tt) \tag{13}$$

s_t = total drawdown with partially penetrating wells in ft

s = drawdown with fully penetrating wells in ft

s_p = drawdown due to effects of partially penetrating wells in ft

Q = production well discharge rate in gpm

r = distance between production and observation wells in ft

S = aquifer storativity as a decimal

T = aquifer transmissivity in gpd/ft

t = pumping period in days

m = aquifer thickness in ft

P_V = aquifer vertical hydraulic conductivity in gpd/sq ft

P_H = aquifer horizontal hydraulic conductivity in gpd/sq ft

l = distance from aquifer top to production well screen base in ft

d = distance from aquifer top to production well screen top in ft

lo = distance from aquifer top to observation well screen base in ft

do = distance from aquifer top to observation well screen top in ft

The partial penetration well function $W[u,r(P_V/P_H)^{1/2}/m, l/m,d/m,lo/m,do/m]$ is calculated with the following equation (Hantush, 1964, pp. 347–358):

$$W[u,r(P_V/P_H)^{1/2}/m,l/m,d/m,lo/m,do/m] = 2m^2/[\pi^2(l-d)(lo-do)] \sum_{n=1}^{50} 1/n^2(\sin n\pi l/m - \sin n\pi d/m)(\sin n\pi lo/m - \sin n\pi do/m)W[u,n\pi r(P_V/P_H)^{1/2}/m] \tag{14}$$

Approximation equations (see Walton, 1985, p. 338) are utilized for estimating values of $W[u,n\,\pi r(P_V/P_H)^{1/2}/m]$. Variables in Equations 11–14 are user-defined in subroutine PARPEN. Production well discharge is assumed constant.

Well Storage Capacity

Similar to PLASM, GWFL3D in algorithms within subroutine SIMEQ assumes well storage capacity is negligible. Whether or not well storage capacity should be simulated may be determined with the following equation (see Driscoll, 1986, p. 566):

$$t_s = 5.4 \times 10^5(r_w^2 - r_c^2)/T \tag{15}$$

where

t_s = time after pumping started beyond which well storage capacity impacts are negligible (less than 1% of drawdown values) in min

r_w = production well effective radius in ft

r_c = pump-column pipe radius in ft

T = aquifer or aquifer layer transmissivity in gpd/ft

Well storage capacity may be simulated in GWFL3D through subroutine WELSTOR with an iterative procedure described by Huyakorn and Pinder (1983, pp. 127–128). The approximate procedure assumes that discharge derived from the aquifer or aquifer layer varies linearly instead of exponentially during the simulation period and uses the principle of superposition. Total well discharge is the sum of discharge derived from the aquifer or aquifer layer and discharge derived from storage within the well bore. Drawdown with well storage impacts is the product of drawdown without well storage impacts and the discharge derived from the aquifer, divided by total well discharge.

The equation relating the total constant discharge from the production well, discharge derived from the aquifer, and discharge derived from storage within the well bore, assuming constant well geometry, is as follows (see Huyakorn and Pinder, 1983, pp. 127–128):

$$Q = Q_{at} + \pi(r_w^2 - r_c^2)s_{ws}7.48/\Delta t \qquad (16)$$

where

Q = constant discharge from production well in gpm

Q_{at} = average discharge derived from aquifer or aquifer layer during simulation period in gpm

r_w = production well effective radius in ft

r_c = pump column-pipe radius in ft

s_{ws} = drawdown with well storage impacts at end of simulation period in ft

Δt = time at end of present simulation period in min

π = 3.141592654

The iterative procedure starts with a first trial value of s_{ws} based on the numerically calculated drawdown without well storage impacts. Equation 16 is used to calculate a first trial value of Q_{at}. A second trial value of s_{ws} is calculated by multiplying drawdown without well storage impacts (s) by the ratio first trial value of Q_{at} divided by total discharge from the well, as shown in the following equation:

$$s_{ws} = sQ_{at}/Q \qquad (17)$$

The first and second trial values of s_{ws} are compared and if the difference exceeds an error tolerance, the iteration is repeated. If the difference is less than or equal to an error tolerance, the estimated values of Q_{at} and s_{ws} are declared valid. The method (successive bisection) involves deciding whether estimated values of s_{ws} are too high or too low in relation to the known value of drawdown without well storage impacts. Variables in Equations 16 and 17 are user-defined in subroutine WELSTOR. Production well dis-

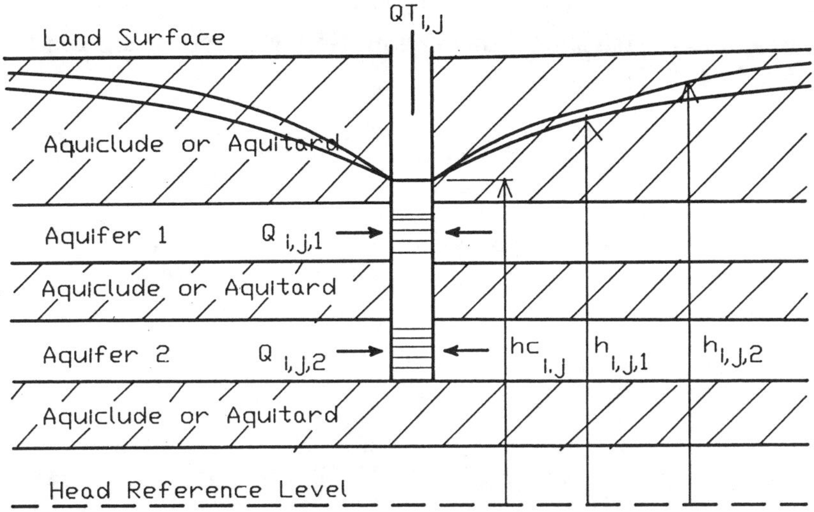

Figure 4. Cross section through a multi-aquifer production well.

charge is assumed to be constant during simulation periods.

Multi-Aquifer Production Well

A multi-aquifer production well may be simulated with an approximate method developed by Bennett et al. (1982, pp. 334–341) through subroutine SIMEQ. The method allows calculation of the water level in the well and individual aquifer or aquifer layer discharges to the well. A non-pumping case is covered in which there may be flow through the well between aquifers or aquifer layers. Negative discharges (recharging well) also may be simulated. A single multi-aquifer production well is assumed to exist in the aquifer system as illustrated in Figure 4.

The discharge rates into the production well from aquifers or aquifer layers open to the well, which appear as $Q_{i,j,k}$ in PLASM (Prickett and Lonnquist, 1971, pp. 46–48), are calculated with the following equation (Bennett et al., 1982, pp. 334–341)

$$Q_{i,j,b} = 2\pi T_{i,j,b} h_{i,j,b}/[\ln(r_e/r_w)] -$$

$$\{2\pi T_{i,j,b}/[\ln(r_e/r_w)]\}\{\sum_{b=m}^{n} [T_{i,j,b} h_{i,j,b}$$

$$/\ln(r_e/r_w)]/\sum_{b=m}^{n} [T_{i,j,b}/\ln(r_e/r_w)]\}$$

$$+ T_{i,j,b} Q T_{i,j}/\{\ln(r_e/r_w)\sum_{b=m}^{n} [T_{i,j,b}/\ln(r_e/r_w)]\} \quad (18)$$

where

$$r_e = 0.28[(T_{yy}/T_{xx})^{1/2}\,\Delta X^2 + (T_{xx}/T_{yy})^{1/2}\,\Delta Y^2]^{1/2} /[(T_{yy}/T_{xx})^{1/4} + (T_{xx}/T_{yy})^{1/4}] \quad (19)$$

$T_{i,j,b}$ = average aquifer or aquifer layer transmissivity at production well node i,j in aquifer or aquifer layer b in gpd/ft

$h_{i,j,b}$ = head in aquifer or aquifer layer b at node i,j at end of time increment in ft

r_e = equivalent production well block radius in ft

T_{xx} = aquifer or aquifer layer b transmissibility in X-direction at well node in ft

T_{yy} = aquifer or aquifer layer b transmissivity in Y-direction at well node in ft

ΔX = well block grid spacing in X-direction in ft

ΔY = well block grid spacing in Y-direction in ft

r_w = production well effective radius in ft

$QT_{i,j}$ = production well discharge rate in gpd

m,n = range of summation which includes each aquifer or aquifer layer open to the production well

$Q_{i,j,b}$ = discharge rate from aquifer or aquifer layer b into well in gpd

Values of $h_{i,j,b}$ are generated by iterative techniques and substituted into the following equation to calculate the composite water level in the production well ($hc_{i,j}$) (Bennett et al., 1982, p. 337):

$$hc_{i,j} = \sum_{b=m}^{n} [T_{i,j,b} h_{i,j,b} / \ln(r_e/r_w)] / \sum_{b=m}^{n} [T_{i,j,b} / \ln(r_e/r_w)] - QT_{i,j} / \{2\pi \sum_{b=m}^{n} [T_{i,j,b} / \ln(r_e/r_w)]\} \quad (20)$$

Then, values of $hc_{i,j}$ and $h_{i,j,b}$ are substituted in the following equation to estimate the rate of flow in gpm from each aquifer or aquifer layer ($Q_{i,j,b}$) into the production well (Bennett et al., 1982, p. 336):

$$Q_{i,j,b} = 2\pi T_{i,j,b} (h_{i,j,b} - hc_{i,j}) / \ln(r_e/r_w) \quad (21)$$

Variables in Equations 18–21 are user-defined in subroutine SIMEQ.

Flowing Well

The flowing well illustrated in Figure 5 may be simulated through subroutine FWEL. Upon removal of the cap from

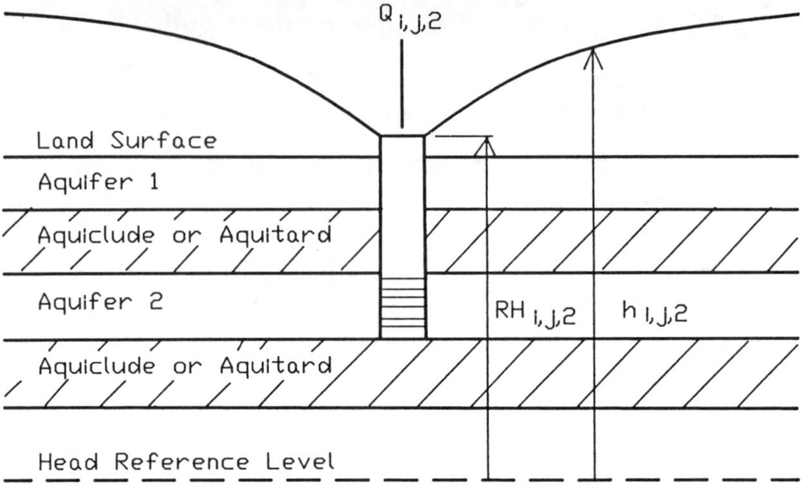

Figure 5. Cross section through a flowing well.

the well, the head in the well declines to the top of the casing and remains constant. Discharge from the well is time-variable and decreases with declining heads in the aquifer or aquifer layer. If the head in the aquifer or aquifer layer declines below the top of the casing, flow ceases. Flowing well simulation is based on the Thiem equation (see Todd, 1980, p. 117) defining steady state flow from a well as follows:

$$Q = 2\pi T(h_2 - h_1)/[\ln(r_2/r_1)] \tag{22}$$

where

Q = flow from well in gpd

T = aquifer or aquifer layer transmissivity in gpd/ft

r_1,r_2 = distances from flowing well to observation points 1 and 2 in ft

h_1,h_2 = heads at observation points 1 and 2 in ft

Nonsteady state flow from the well can be simulated numerically by applying the Thiem equation to successive time increments. h_1 may be assumed to be the head ($RH_{i,j,k}$) at the flowing well with r_1 equal to the effective flowing well radius (r_w). h_2 is assumed to be the head ($h_{i,j,k}$) calculated at the flowing well node with r_2 equal to the equivalent flowing well block radius (r_e). With rectangular grid spacing and anisotropic conditions, equation 22 may be rewritten as (Peaceman, 1983):

$$Q = 2\pi(T_{xx}T_{yy})^{1/2}(h_2 - h_1)/[\ln(r_e/r_w)] \quad (23)$$

where

$$r_e = 0.28[(T_{yy}/T_{xx})^{1/2}\,\Delta X^2 + (T_{xx}/T_{yy})^{1/2}\,\Delta Y^2]^{1/2} \\ /[(T_{yy}/T_{xx})^{1/4} + (T_{xx}/T_{yy})^{1/4}] \quad (24)$$

T_{xx} = aquifer or aquifer layer transmissivity in X-direction at well node in ft

T_{yy} = aquifer or aquifer layer transmissivity in Y-direction at well node in ft

ΔX = well block grid spacing in X-direction in ft

ΔY = well block grid spacing in Y-direction in ft

Substitution of $RH_{i,j,k}$, r_w, $h_{i,j,k}$, and r_e in Equation 23 results in the expression:

$$Q_{i,j,k} = \{2\pi(T_{xx}T_{yy})^{1/2}/[\ln(r_e/r_w)]\}(h_{i,j,k} - RH_{i,j,k}) \quad (25)$$

A flowing well simulation factor ($RF_{i,j,k}$) is assigned to the flowing well node and is defined as:

$$RF_{i,j,k} = 2\pi(T_{xx}T_{yy})^{1/2}/[\ln(r_o/r_w)] \quad (26)$$

Flow from the well is given by the following equation:

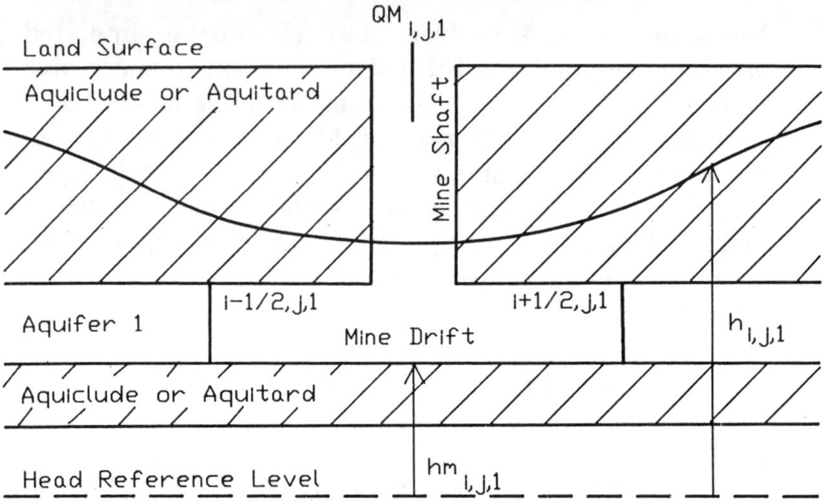

Figure 6. Cross section through a mine.

$$Q_{i,j,k} = RF_{i,j,k}h_{i,j,k} - RF_{i,j,k}RH_{i,j,k} \tag{27}$$

All flowing wells in the system are capped simultaneously upon request. Variables in Equations 22–27 are user-defined in subroutine FWEL.

Mines

A mine shaft and drift illustrated in Figure 6 may be simulated as a constant head-variable discharge drain (see OSM, 1981, pp. A21–A-22) through subroutine MINE. The shaft has sealed walls and is centered at the mine node. The drift cavity extends the full aquifer or aquifer layer thickness to distances midway between adjacent nodes. The grid at and near the mine node is square. The drift-aquifer face is uniform and perpendicular to the aquifer base. The drift intercepts all flow toward the mine node. At least two nodes separate individual mines and any aquifer system boundaries.

An unsealed shaft may be simulated with the flowing well simulation techniques by replacing the top of the casing elevation with the drift floor elevation. Irregular mine drifts may be simulated by appropriate creation of adjacent mine nodes.

A constant head is created at the mine node by setting the storage factor at 1E21. The mine floor elevation $hm_{i,j,k}$ is user-defined in subroutine MINE. The head in the aquifer is not permitted to decline below the drift floor. A form of Darcy's law useful in defining node-to-node flow rates is (Ferris et al., 1962, p. 73):

$$Q_d = TIL \tag{28}$$

where

$$I = \Delta h / D_f \tag{29}$$

Q_d = flow rate through an aquifer or aquifer layer cross section in gpd

T = aquifer or aquifer layer cross section transmissivity in gpd/ft

I = hydraulic gradient in ft/ft

Δh = head loss between nodes in ft

L = width of aquifer or aquifer layer cross section through which flow occurs in ft

D_f = flow distance in ft

Similar to PLASM, flow distances in GWFL3D are equal to the grid spacing, whereas at mine nodes, flow distances are equal to one-half of the grid spacing because the drift-aquifer face is midway between nodes. Since flow rates at all nodes are based on head losses and distances between nodes (grid spacing), T at mine nodes must be divided by a mine node transmissivity adjustment factor (T_m) equal to

0.5 to compensate for the drift-aquifer face position and satisfy Equation 29.

Mine discharge from the four sides of the rectangular cavity calculated in subroutine WATBL and based on Equation 28 is:

$$QM_{i,j,k} = T2_{i-1,j,k}(h_{i-1,j,k} - h_{i,j,k}) + T2_{i,j,k}(h_{i+1,j,k} - h_{i,j,k}) + T1_{i,j,k}(h_{i,j+1,k} - h_{i,j,k}) + T1_{i,j-1,k}(h_{i,j-1,k} - h_{i,j,k}) \quad (30)$$

where

$QM_{i,j,k}$ = mine discharge in gpd

$T1_{i,j,k}$ = aquifer or aquifer layer transmissivity within the vector volume between nodes i,j and i,j+1 in gpd/ft (adjusted for mine cavity)

$T2_{i,j,k}$ = aquifer or aquifer layer transmissivity within the vector volume between nodes i,j and i+1,j in gpd/ft (adjusted for mine cavity)

$h_{i,j,k}$ = mine floor elevation at node i,j in ft

Drainage of the mine cavity is not included in numerical calculations and must be estimated with the following equation and added to $QM_{i,j,k}$ during the mine drift excavation time increment:

$$QD_{i,j,k} = 7.48m\ \Delta X\ \Delta Y S_y/\Delta t_{md} \quad (31)$$

where

m = aquifer or aquifer layer thickness in ft

ΔX = grid spacing in X-direction in ft

ΔY = grid spacing in Y-direction in ft

Table 1. Values of T_m as a Function of Mine Penetration and P_H/P_V Ratio

Percent Mine Penetration	P_H/P_V Ratio				
	1	2	10	50	100
20	.516	.530	.583	.715	.836
40	.507	.514	.548	.612	.667
60	.504	.506	.519	.549	.572
80	.501	.501	.505	.516	.523
100	.500	.500	.500	.500	.500

S_y = aquifer or aquifer layer specific yield as a decimal

Δt_{md} = mine drift excavation time period in days

A partially penetrating mine drift similar to that illustrated in Figure 6 may be simulated by further adjustment of T_m based on the degree of mine drift penetration into the aquifer and the ratio of the horizontal and vertical hydraulic conductivities (P_H/P_V) of the aquifer (Prickett, personal communication, 1983). Flow into the mine drift decreases as the degree of penetration decreases and the P_H/P_V ratio increases because of variable vertical head losses in addition to horizontal head losses with full penetration.

A separate cross-sectional aquifer slice model (see Walton, 1985, p. 391), based on PLASM, was utilized to calculate the values of T_m under selected partial penetration conditions in Table 1. T_m is the ratio of the flow into a mine node versus the flow into a non-mine node with a constant head loss and selected mine penetrations and conductivity ratios.

The depth of penetration into the aquifer is substituted into Equation 31 instead of m to calculate mine drift cavity drainage under partial penetrating conditions.

Mine drift stepwise advancement may be simulated by creating constant heads sequentially in time at adjacent mine nodes. Small time increments should be specified each time there is a change in the mine schedule. After mining is

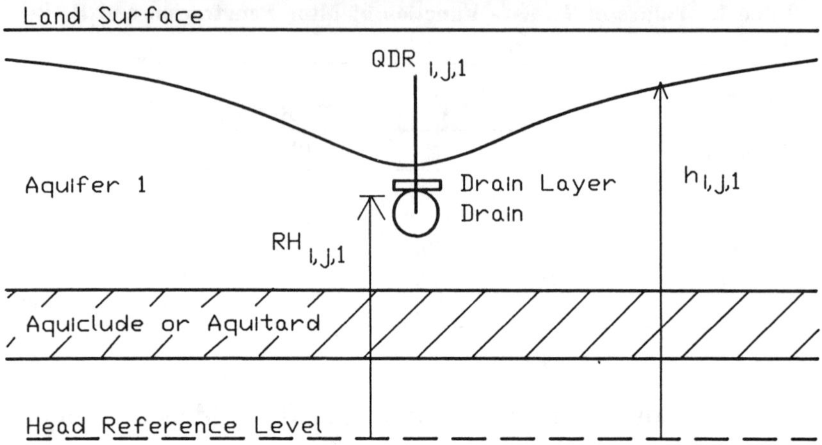

Figure 7. Cross section through a drain.

complete, constant heads are released and T_m is reset to 1.0. It is assumed that the mine drift cavities are backfilled with spoil materials. The hydraulic conductivity and storativity of the spoil materials may be user-defined in subroutine MINE, together with reclamation times.

Drains

Seepage into the drain illustrated in Figure 7 may be simulated in a manner analogous to the simulation of induced infiltration of surface water through a streambed in subroutine DRAIN. The head in the drain is assumed to be the elevation of the drain top $RH_{i,j,1}$. A drain layer, materials between the drain and aquifer material, represents drain tile openings with any chemical precipitation and the backfill around the tile. The effective horizontal width of the drain layer is assumed to be the drain tile diameter plus the drain layer thickness multiplied by two. It is further assumed that the aquifer head does not decline below the drain top.

According to Darcy's law, the drain discharge rate is directly proportional to the drain horizontal area (effective horizontal width of drain layer multiplied by drain length), the vertical hydraulic conductivity of the drain layer, and the head difference between that in the aquifer or aquifer layer and in the drain; it is inversely proportional to the drain layer thickness as indicated in the following equation (see McDonald and Harbaugh, 1984, pp. 288–292):

$$Q_d = (P_d/m_d)\Delta h A_d \tag{32}$$

where

$$A_d = W_d L_d \tag{33}$$

Q_d = drain discharge in gpd

P_d = vertical hydraulic conductivity of drain layer in gpd/sq ft

m_d = drain layer thickness in ft

Δh = difference between head in aquifer or aquifer layer and head in drain in ft

W_d = effective drain layer width in ft

L_d = length of drain assigned to node in ft

A drain simulation factor, in gpd/ft, is assigned to all drain nodes in the uppermost aquifer or aquifer layer and is defined as:

$$RN_{i,j,1} = (P_d/m_d)A_d \tag{34}$$

Equation 32 can be rewritten in terms of $RN_{i,j,1}$ and the head notation in Figure 7 as:

$$QDR_{i,j,1} = RN_{i,j,1}h_{i,j,1} - RN_{i,j,1}RH_{i,j,1} \tag{35}$$

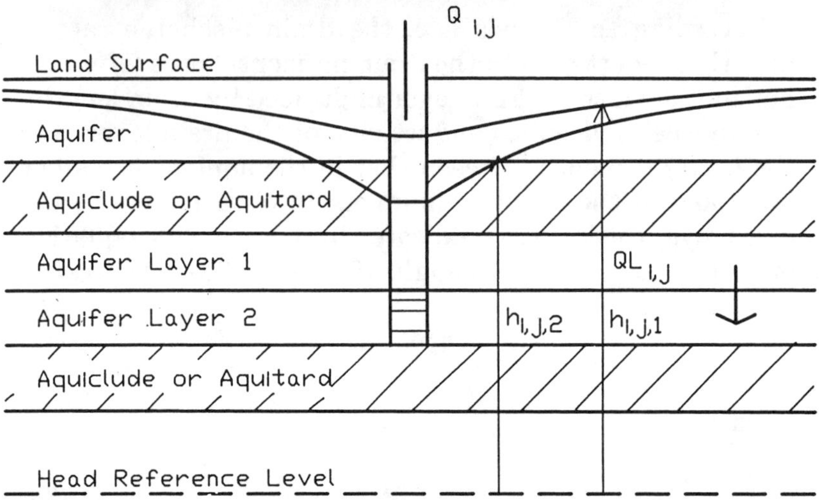

Figure 8. Cross section through a layered aquifer.

Variables in Equations 32–35 are user-defined in subroutine DRAIN.

Aquifer Layering

The layered aquifer illustrated in Figure 8 may be simulated by assigning to each node where layering occurs an aquifer layering simulation factor ($RA_{i,j}$) in gpd/ft defined as follows (see McDonald and Harbaugh, 1984, pp. 130, 134, 142):

$$RA_{i,j} = C_1C_2/(C_1 + C_2) \tag{36}$$

where

$$C_1 = PV_{i,j,1}AI_{i,j}/(m_{i,j,1}/2) \tag{37}$$

$$C_2 = PV_{i,j,2}AI_{i,j}/(m_{i,j,2}/2) \tag{38}$$

$PV_{i,j,1}$ = aquifer layer 1 vertical hydraulic conductivity in gpd/sq ft

$PV_{i,j,2}$ = aquifer layer 2 vertical hydraulic conductivity in gpd/sq ft

$AI_{i,j}$ = horizontal area of inter-aquifer flow transfer centered at node in sq ft

$m_{i,j,1}$ = aquifer layer 1 thickness in ft

$m_{i,j,2}$ = aquifer layer 2 thickness in ft

The rate of flow between aquifer layers ($QL_{i,j}$) in gpd is as follows (McDonald and Harbaugh, 1984, p. 130):

$$QL_{i,j} = RA_{i,j}(h_{i,j,1} - h_{i,j,2}) \tag{39}$$

where

$h_{i,j,1}$ = head in aquifer layer 1 in ft

$h_{i,j,2}$ = head in aquifer layer 2 in ft

The rate of flow reaches a maximum value and remains constant thereafter when water levels in aquifer layer 2 decline below the top of that layer. Storativity in an aquifer layer is artesian, unless the head in the layer is less than the layer top elevation; then it is water table.

Aquitard Storativity

The time period during which aquitard storativity impacts are appreciable and should be simulated can be calculated with the following equation (Bredehoeft and Pinder, 1970, pp. 887–888; Marsily, 1986, p. 363):

$$t_{as} = S'm'/P' \tag{40}$$

where

t_{as} = time after last pumping rate change during which release of water stored in aquitard is appreciable (simulation of aquitard storativity is important) in days

S' = aquitard storativity as a decimal

m' = aquitard thickness in ft

P' = aquitard hydraulic conductivity in gpd/sq ft

Release of water stored in an aquitard may be simulated in GWFL3D during periods when time is less than t_{as} by assigning to each node where the aquitard exists, an aquitard storativity simulation factor defined as (Trescott et al., 1976, pp. 4–7; Bredehoeft and Pinder, 1970, pp. 883–888):

$$RT_{i,j,k} = P'_{i,j,k}A_{i,j,k}/\{[\pi P'_{i,j,k}t_{ap}m'_{i,j,k}/(3m'^2_{i,j,k}S'_{i,j,k}7.48)]^{1/2}m'_{i,j,k}\}\{1 + 2\sum_{n=1}^{200} \exp\{-n^2/[P'_{i,j,k}t_{ap}m'_{i,j,k}/(3m'^2_{i,j,k}S'_{i,j,k}7.48)]\}\} \tag{41}$$

where

$RT_{i,j,k}$ = aquitard storativity simulation factor in gpd/ft

$A_{i,j,k}$ = horizontal area of aquitard vector volume within which water is released from storage in sq ft

$S'_{i,j,k}$ = aquitard storativity as a decimal

t_{ap} = time after last pumping rate change in days

It is assumed that transient leaky effects from previous

pumping periods have dissipated (each pumping rate period exceeds t_{as} in the case of multi-pumping rate schedules). The aquitard storage release rate ($QS_{i,j,k}$) in gpd is as follows:

$$QS_{i,j,k} = RT_{i,j,k}ho_{i,j,k} - RT_{i,j,k}h_{i,j,k} \qquad (42)$$

where

$h_{i,j,k}$ = head in aquifer or aquifer layer at time t in ft

$ho_{i,j,k}$ = head in aquifer or aquifer layer at end of previous time increment in ft

Delayed Gravity Yield

The time period during which delayed gravity yield under water table conditions should be simulated may be determined with the following equation (Walton, 1962, p. 6):

$$t_d = 37.5mS_y/P_V \qquad (43)$$

where

t_d = time after pumping started beyond which delayed gravity yield impacts are negligible in days

m = aquifer or aquifer layer thickness in ft

S_y = aquifer or aquifer layer specific yield as a decimal

P_V = aquifer or aquifer layer vertical hydraulic conductivity in gpd/sq ft

Delayed gravity yield may be simulated in GWFL3D, assuming a single aquifer or aquifer layer in the aquifer system by assigning to each node an effective storativity and an effective recharge defined as (Rushton and Redshaw, 1979, pp. 259–262):

$$SE_{i,j,k} = S_{i,j,1} + SY_{i,j,1}[1 - \exp(-alpha_{i,j,1}\ \Delta t_t)] \tag{44}$$

$$QR_{i,j,1} = alpha_{i,j,1} SY_{i,j,1} \sum_{i=1}^{n-1} \Delta s_m \exp[-alpha_{i,j,1} (t_n - t_m)] 7.48 A_{i,j,1} \tag{45}$$

where

$alpha_{i,j,l}$ = $3PV_{i,j,l}/(SY_{i,j,1} m 7.48)$ (46)
(See Neuman, 1979, pp. 899–908; Streltsova, 1972, pp. 1059–1066)

$SE_{i,j,1}$ = effective aquifer or aquifer layer storativity with delayed gravity yield during current time increment as a decimal

$S_{i,j,1}$ = aquifer or aquifer layer storativity as a decimal

$SY_{i,j,1}$ = aquifer or aquifer layer specific yield as a decimal

$alpha_{i,j,1}$ = reciprocal of delayed index in day $^{-1}$

Δt_t = time after pumping started in days

Δs_m = drawdown during the m th time increment in ft

t_m = time at the end of the m th time increment in days

t_n = time at the end of the n th time increment in days

n = current time increment number

$QR_{i,j,1}$ = effective recharge arising from delayed gravity yield due to previous drawdown in gpd (during m th time increment, a drawdown s_m occurred which produces a delayed gravity yield contribution at a later time t_n)

$A_{i,j,1}$ = horizontal area of aquifer or aquifer layer vector volume in which delayed gravity yield occurs in sq ft

$PV_{i,j,1}$ = aquifer or aquifer layer vertical hydraulic conductivity in gpd/sq ft

$h_{i,j,1}$ = head in aquifer or aquifer layer at end of current time increment in ft

Small time increments (minutes or less) and grid spacings near the production well (usually 10's of feet) and a small nodal head convergence error (usually 0.001) are required for precise simulation of delayed gravity yield. The simulation algorithm assumes there is one production well and no flowing wells, drains, mines, groundwater evapotranspiration, or areal recharge.

4

Contaminant Transport Program Operation

Contaminant transport with or without adsorption and radioactive decay during one simulation time increment and in one aquifer or aquifer layer is simulated in GWTR3D. Contaminant transport during more than one simulation time increment and/or more than one aquifer or aquifer layer may be simulated with multiple program runs.

Variables are dimensioned in COMMON statements and defined in subroutine INFOM. Display of operation information at intervals during the program is optional. Only English gallon-day-foot system of units is supported for groundwater flow. Milligram-liter-pound units are supported for contaminant transport. Required units are specified in interactive data base input statements. Useful unit conversions are: 1 gram = 2.205×10^{-3} pounds and 1 cu meter = 2.642×10^{2} US gal.

The user may choose importation of a data base through subroutine FILEI1 or keyboard interactive data base input. Pre-existing contaminant conditions may be simulated with user keyboard interactive specification of nodal concentrations at the start of the simulation time increment or importation of a file through subroutine FILEI3.

The finite-difference grid (number of columns and rows, and default grid spacing) are user-defined. In GWTR3D, the number of columns or rows must be less than 51. A hard copy of grid spacings is optional. The user may specify variable grid spacings within selected grid blocks. The duration of the simulation time increment is user-defined.

The user fills the following variable arrays with default values to simulate homogeneous conditions: aquifer or aquifer layer horizontal hydraulic conductivity, top elevation, base elevation, and actual and effective porosity. Other variables are initialized by calling subroutine INIT. The user may revise variable values, except those for porosity, within selected grid blocks to simulate heterogeneous conditions.

Source and sink geometry and strength, and aquifer system hydraulic characteristics at sources and sinks, are user-defined in subroutine CONSTR. This subroutine covers: pre-existing contaminant concentration at nodes which is user-defined at the keyboard and stored in a file through subroutine FILEO3 or imported through subroutine FILEI3; number of slug sources; particle mass; maximum number of particles allowed in model (5000); aquifer distribution coefficient; aquifer bulk mass density; calculation of aquifer or aquifer layer retardation factor; half-life of radioactive isotope; number of injection wells and injection well coordinates; contaminant leakage through aquitards with keyboard user-defined aquitard vertical hydraulic conductivity, thickness, and sourcebed concentration, sourcebed heads which may be user-defined at the keyboard and stored in a file or imported from a file, and calculation of aquitard leakance; number of local source pits or basins, vertical hydraulic conductivity of source pit or basin bed, thickness of source pit or basin, and calculation of source pit or basin leakance; number of stream source or sink nodes, stream source or sink node coordinates, streambed vertical hydraulic conductivity, streambed thickness, stream surface elevation, streambed base elevation, streambed width and length, and calculation of stream source or sink simulation factor; number of production well sinks, production well sink coordinates, and production well sink discharge rate; areal sink aquitard vertical hydraulic conductivity and thickness, areal sink sourcebed head which may be user-defined at the keyboard and stored in a file or imported from a file, areal sink aquifer layer vertical hydraulic conductivity and thickness, areal sink aquifer layer heads user-defined at the keyboard and stored in a file or imported from a file, areal sink aquitard distribution coefficient, areal sink bulk mass density, and calculation of areal sink retardation factor.

The user may choose to store the data base created hereto in files through subroutine FILEO1 and obtain a hard copy of the data base through subroutine DATPR. If the data base was imported, the keyboard data base entry would be bypassed and the program would continue at this point.

The user may choose interactive keyboard aquifer or aquifer layer head data base specification or importation of a head data base file through subroutine FILEI2. If keyboard specification is selected, the user may create a head data base file through subroutine FILEO2. The user may choose to create a uniform gradient in the X-direction head data base through subroutine HEAD by specifying the water level elevation along column 1 and the uniform groundwater seepage velocity. The head data base should reflect water level impacts due to any simulated contaminant source or sink. Internodal velocities are calculated (see Prickett et al., 1981, pp. 7,21). Longitudinal and transverse dispersivities may be estimated as a function of the mean contaminant migration distance during the simulation time increment. The user may request values of groundwater grid velocities at selected nodes to assist in the estimation of dispersivities through subroutine DISPER. Dispersivities and velocities are adjusted for any adsorption with a retardation factor user-defined in subroutine CONSTR.

Depending upon contaminant conditions specified, some or all of the following subroutines are called: CLEAR, SLUGIN, PRESOR, ADVAN, ARESOR, CONTIN, INJIN, LOCSOR, ARESNK, and AREA3D. As a result, large numbers of particles representing contaminant mass may be introduced at user-defined source locations with I,J coordinates assigned in subroutine ADD called for in slug source subroutines SLUGIN and PRESOR. In function Move, called for in subroutine ADVAN, each particle is moved (particle coordinates are changed) with the velocity of flow to simulate advection of contaminants. Each particle is then again moved in a random direction an additional distance to simulate dispersion. The random movement is governed by scaled probability curves related to flow length and dispersion coefficients (see Bear and Verruijt, 1987, pp. 336–339).

Subroutine ADD and function MOVE are also called for in continuous source and areal sink subroutines ARESOR, CONTIN, INJIN, LOCSOR, ARESNK, and AREA3D. Function RAN is called for in continuous source and sink subroutines to distribute particles randomly in space and time and in function ANORM to generate random movement. Function VIJK is called for in subroutine VELO, and subroutine VELO and function ANORM are called for in function MOVE.

Details concerning subroutines ADD, ADVAN, CLEAR, INIT, and VELO and functions ANORM, VIJK, and MOVE are provided by Prickett et al. (1981, pp. 53–63). In general, ADD assigns I,J coordinates to source particles; ADVAN, through MOVE, advances all slug source particles to new positions; ANORM produces a random number between -6 and 6 with normal distribution based on a random number between 0 and 1 generated by RAN; CLEAR clears tabulations of number of particles assigned to nodes or captured by sinks; MOVE moves particles to new positions, tabulates in which nodal area particles reside, and tabulates any particles captured by sinks; VIJK provides proper velocities to VELO; and VELO interpolates block velocities from grid velocities for particle movement within grid blocks.

In function MOVE, particles are kept within grid boundaries which extend one-half grid spacing beyond outer grid lines. If the contaminant plume extends beyond outer grid lines, particles with positions outside outer grid lines will not be tabulated as being within the grid boundaries. The particle balance (particles in through sources minus particles out through sinks = 0) will be in error. In this case, adjustments in grid spacing and/or simulation time increment should be made and the program rerun so that the plume does not extend beyond outer grid lines.

Quasi 3-D areal vertical contaminant transport is simulated through areal sink and source subroutines. For example, during program run 1, areal vertical leakage of contaminants from aquifer 1 through an aquitard and into aquifer 2 may be simulated as areal sink mass in subroutine ARESNK. During program run 2, this areal sink mass becomes areal source mass in aquifer 2 through subroutine AREA3D. Thus, quasi 3-D contaminant transport may be simulated with multiple program runs.

Particles which migrate to positions within one-half grid spacing near sink positions are captured by the sink. Other particles are assigned to the nearest nodes. The concentration at each node is determined by dividing the sum of the mass of the particles assigned to nodes by the volume of water at the node (see Prickett et al., 1981, p. 50). Sources and sinks are assumed to fully penetrate the aquifer or aquifer layer.

Screen displays of nodal concentrations and particle distribution are optional. The user may specify that hard copies be made by a printer of nodal and sink concentrations and particle distributions through subroutines CONMAP, SUMMRY, and SNKCON (see Prickett et al., 1981, pp. 69–70). Contamination concentration is divided by the retardation factor to simulate contamination with adsorption. This reflects the fact that part of the contaminant mass is adsorbed onto the aquifer skeleton and does not contribute to dissolved concentration. To simulate decontamination with desorption, contaminant concentration is multiplied by the retardation factor, thus reflecting the fact that contaminant mass is desorbed from the aquifer skeleton and contributes to dissolved concentration. Capture wells are assumed to be located within the contaminant plume in desorption simulation. Desorption distribution coefficients should reflect the fact that sorption processes are not completely reversible. The user may choose to create a file of calculated concentration nodal results, convert this file into a file formatted for export to commercial graphics programs through subroutine XYZDTC, and display a profile view of nodal concentrations in multi-aquifer or aquifer layer situations through subroutine PROFIL.

5

Contaminant Transport Simulation Techniques

Most contaminant transport simulation techniques utilized in GWTR3D are described in detail by Prickett et al. (1981). Selected details are repeated herein to assist the reader in understanding modifications and extensions to RANDOM WALK. Simulation techniques not covered in RANDOM WALK are emphasized in the following sections of this book.

Grid Orientation

The model grid lines superposed over the aquifer are indexed using the I (column), J (row) notation colinear with the X and Y directions, respectively. I-coordinates increase left to right and J-coordinates increase top to bottom. Grid spacing is centered at the nodes (grid line intersections). The origin is the upper-left corner. Contaminants migrate on the basis of an X,Y coordinate system congruent with and on the same scale as the I,J grid. Grid boundaries must extend beyond the contaminant plume. The grid density should be greatest at sources and at potential impact locations such as sinks. The number of columns or rows must match the number of columns or rows used in creating any head data file to be imported.

Pre-Existing Conditions

Subroutine PRESOR generates particles at the beginning of the simulation time increment to simulate pre-existing contaminant conditions. Total pre-existing contaminant load at each node is calculated with the following equation (Prickett et al., 1981, p. 50):

$$PL_p = C_n \, \Delta X \, \Delta Ym \text{ APOR } 62.4/1E6 \tag{47}$$

where

PL_p = total nodal pre-existing contaminant load at node in pounds

C_n = contaminant concentration at node in mg/L

ΔX = grid spacing in X-direction at node in ft

ΔY = grid spacing in Y-direction at node in ft

m = aquifer or aquifer layer thickness in ft

APOR = aquifer or aquifer layer actual porosity as a decimal

note: 62.4 pounds per cubic feet of water

The model grid and any pre-existing contaminant loads must extend beyond the area under investigation to allow migration of pre-existing particles to and from that area. Contaminant concentration at nodes is user-defined in subroutine CONSTR, and the grid spacings and aquifer or aquifer layer thickness and actual porosity are user-defined in the MAIN program.

Particles are generated randomly within areas centered at nodes and extending midway to surrounding nodes until accumulated individual particle mass equals or exceeds the total pre-existing contaminant load. The subroutine ADD

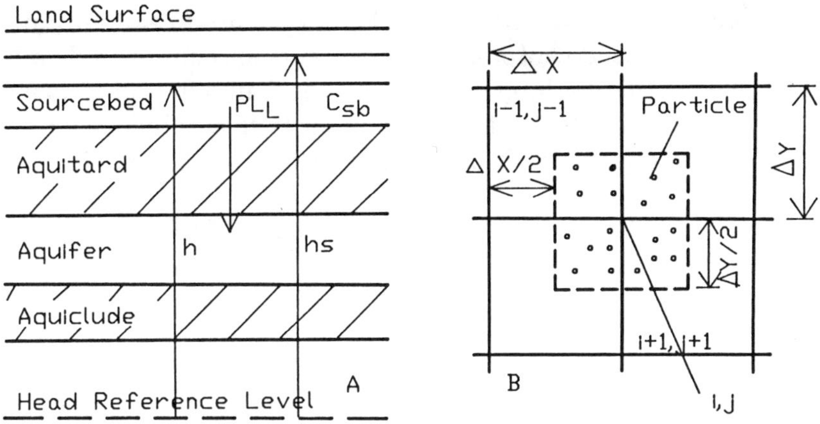

Figure 9. Cross section (A) and plan view (B) through a continuous areal source.

is called to store initial particle coordinates in X and Y arrays.

Areal Sources

Two types of areal sources may be simulated. In subroutine ARESOR, contaminant concentration in a sourcebed above an aquifer or aquifer layer is user-defined and contaminant mass in the sourcebed is not depleted as the result of downward contaminant leakage. In subroutine AREA3D, leakage contaminant mass calculated in subroutine ARESNK with aquifer or aquifer layer contaminant mass depletion as the result of vertical contaminant leakage, constitutes the source.

Subroutine ARESOR continuously generates particles to simulate nodal sources due to contaminant leakage from a sourcebed through an aquitard and into an aquifer or aquifer layer, as shown in Figure 9. It is assumed that the sourcebed head, hydraulic characteristics, and concentration are variable in space and time and the aquifer head is above the aquitard top. Particles are located randomly in time and space centered at nodes and extending midway between surrounding nodes. The total nodal source load during the simulation time increment is calculated with the following equation (Prickett et al., 1981, pp. 51, 52, 70):

$$PL_L = (h_s - h)RL\ \Delta t\ C_{sb}\ 8.3453/1E6 \tag{48}$$

where

$$RL = (P'/m')\ \Delta X\ \Delta Y \tag{49}$$

PL_L = total nodal source load during simulation time increment due to leakage of contaminants through aquitard in pounds

h_s = head in sourcebed in ft

h = head in aquifer or aquifer layer in ft

P' = vertical hydraulic conductivity of aquitard in gpd/sq ft

m' = aquitard thickness in ft

ΔX = grid spacing in X-direction at node in ft

ΔY = grid spacing in Y-direction at node in ft

Δt = simulation time increment in days

C_{sb} = concentration in sourcebed in mg/L

The data base is user-defined in subroutine CONSTR and the MAIN program. Sourcebed and aquifer or aquifer layer head data bases may be imported from a file, or user-defined from the keyboard and later exported to a file. Coordinates of grid blocks for selected nodal data base specification are user-defined. Particles are generated until accumulated individual particle mass equals or exceeds the total source load. Each particle's initial coordinates are calculated, the subroutine MOVE is called, and each particle is moved to a new position with new coordinates. If the particle does not enter a sink, the subroutine ADD is called to store particle coordinates in X and Y arrays.

Subroutine AREA3D simulates continuous migration of contaminants from one aquifer through an aquitard and into another aquifer, or between adjacent aquifer layers. Particles at areal sinks centered at nodes identified in subroutine ARESNK are regenerated randomly in time and in spaces centered at corresponding nodes and extending midway between surrounding nodes. The total sink load due to this migration during the simulation time increment is calculated, displayed, and stored on a file in subroutine ARESNK. This file may be imported or the load may be user-defined from the keyboard and becomes the source load. Particles are generated until accumulated individual particle mass equals or exceeds the total source load. Particle mass should be as small as possible to ensure reasonable precision. Each particle's initial coordinates are calculated, the subroutine MOVE is called, and each particle is moved to a new position with new coordinates. If the particle does not enter a sink, the subroutine ADD is called to store particle coordinates in X and Y arrays. This simulation is approximate because of the particle regeneration technique (particle locations are not transferred to next-time increment), and its accuracy may not be sufficient under certain conditions. In this case, the user should not use GWTR3D but use a more sophisticated program (contact IGWMC for information concerning availability of 3-D mass transport programs; see Uffink, 1985, pp. 108–109).

Local Slug Sources

Subroutine SLUGIN simulates slug sources by generating any number of particles uniformly distributed at a point, along a line, around the circumference of a curve, or within a rectangular area at the beginning of the simulation

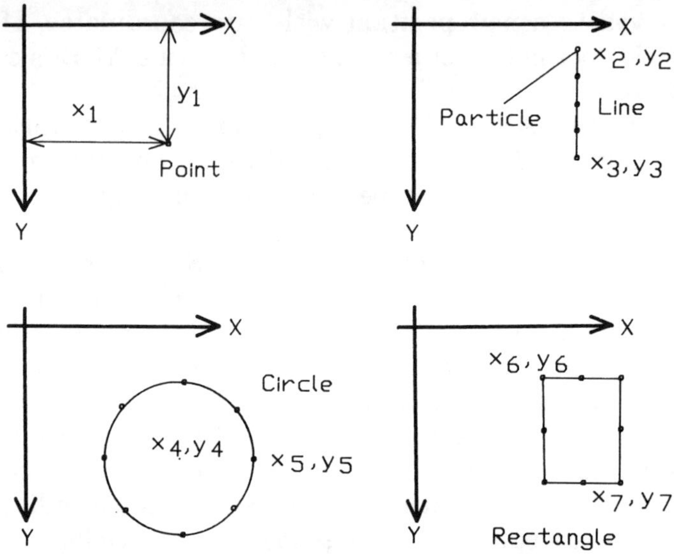

Figure 10. Plan view of slug sources.

time increment, as shown in Figure 10. Necessary source space coordinates (X_1, Y_1 for point; X_2,Y_2 and X_3,Y_3 for the beginning and end of line; X_4,Y_4 and X_5,Y_5 for the center and a point on circumference of circle; X_6,Y_6 and X_7,Y_7 for the corners of rectangle) and the total contaminant slug load are user-defined in this subroutine. The total slug load is divided by the particle mass and the nearest whole number of particles associated with the load is determined. Particle coordinates are calculated on the basis of source space coordinates. The subroutine ADD is called to store initial particle coordinates in X and Y arrays. Subroutine SLUGIN is similar to the RANDOM WALK subroutines GENP2, GENP3, GENP4, and GENP5.

Local Continuous Sources

Two types of local continuous sources may be simulated. In subroutine CONTIN, the total contaminant load is user-defined and is injected directly into the aquifer or aquifer layer, whereas in subroutine LOCSOR, the leachate concentration is user-defined and there is leachate leakage through a sourcebed and into the aquifer or aquifer layer.

Subroutine CONTIN is essentially the GENP subroutine in RANDOM WALK. Continuous sources (landfills, pits, basins, ditches, and surface water bodies) are simulated by generating particles randomly within a user-defined number of rectangular areas, and randomly during the simulation time increment with user-defined contaminant loads. Source area corner coordinates and the total contaminant load reaching the aquifer or aquifer layer during the simulation time increment are user-defined. Point, line, or square source areas may be simulated by proper choice of corner coordinates. Particles are generated randomly in time until accumulated individual particle mass equals or exceeds the total source load. Each particle's initial coordinates within the source area are calculated, the subroutine MOVE is called, and each particle is moved to a new position with new coordinates. If the particle does not enter a sink, the subroutine ADD is called to store particle coordinates in X and Y arrays.

Subroutine LOCSOR simulates continuous migration of contaminants through local sourcebeds (landfills, pits, basins, and surface water bodies) into aquifer or aquifer layers, as shown in Figure 11. Source area corner coordinates (X_1,Y_1 and X_2,Y_2), source concentration, and elevation of aquifer head at the source are user-defined. The sourcebed vertical hydraulic conductivity and thickness are also user-defined in the subroutine CONSTR. The total source contaminant load during the simulation time increment is calculated with the following equation (see Prickett et al., 1981, pp. 51, 52, 70):

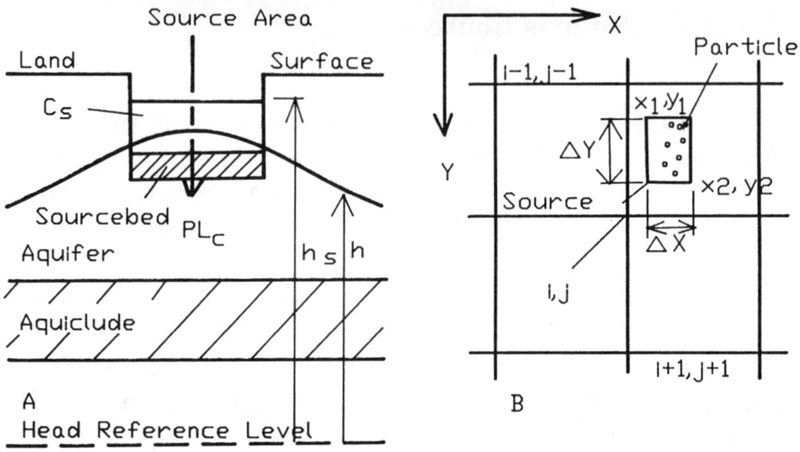

Figure 11. Cross section (A) and plan view (B) through a continuous local source.

$$PL_c = (h_s - h)RC\ \Delta t\ C_s\ 8.3453/1E6 \tag{50}$$

where

$$RC = (P'/m')A_s \tag{51}$$

PL_c = total local contaminant source load in pounds

h_s = average elevation of source water surface in ft

h = average elevation of aquifer or aquifer layer head at source in ft

P' = vertical hydraulic conductivity of sourcebed in gpd/sq ft

m' = local sourcebed thickness in ft

A_s = local sourcebed horizontal area in sq ft

Δt = simulation time increment in days

C_s = source leachate concentration in mg/L

note: 8.3453 lbs/gal
parts per million=mg/L

Particles are generated randomly in the source area space and in time until accumulated individual particle mass equals or exceeds the total source load. Each particle's initial coordinates within the source area are calculated randomly in space, subroutine MOVE is called, and each particle is moved to a new position with new coordinates. If the particle does not enter a sink, subroutine ADD is called to store particle coordinates in X and Y arrays. This subroutine is similar to subroutine SORGEN in RANDOM WALK.

Injection Well Continuous Source

Subroutine INJIN generates particles at selected nodes to simulate continuous injection well contamination. The concentration of the injected water and the well injection rate are user-defined. The total nodal injection well source load during the simulation time increment is calculated with the following equation (Prickett et al., 1981, p. 70):

$$PL_I = Q_I \, \Delta t \, C_I \, 8.3453/1E6 \tag{52}$$

where

PL_I = total nodal source load during simulation time increment in pounds

Q_I = well injection rate in gpd

Δt = simulation time increment in days

C_I = concentration of injected water in mg/L

Simulation time is user-defined in the MAIN program and

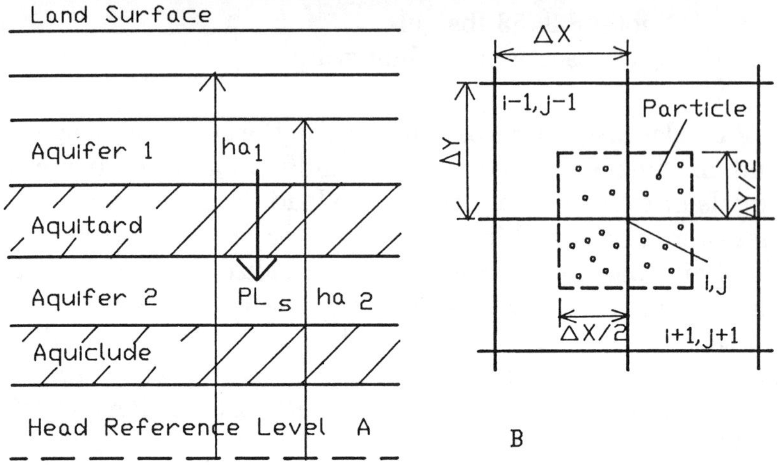

Figure 12. Cross section (A) and plan view (B) through a continuous areal sink.

injection well coordinates are user-defined in subroutine CONSTR.

Particles are generated randomly in time at injection well node coordinates until accumulated individual particle mass equals or exceeds the total contaminant load. Subroutine MOVE is called and each particle is moved to a new position with new coordinates. If the particle does not enter a sink, subroutine ADD is called to store particle coordinates in X and Y arrays.

Areal Sinks

Subroutine ARESNK simulates quasi-3-D vertical migration of contaminants with or without adsorption from one aquifer through an aquitard and into another aquifer, or between adjacent aquifer layers as shown in Figure 12, by creating areal sinks at appropriate nodes. The total nodal areal sink contaminant load is calculated with the following mass balance equation, provided the contaminant load leaving the aquifer unit or layer is less than the contaminant load injected into the aquifer unit or layer:

$$PL_a = PL_t - PL_s \tag{53}$$

where

$$PL_a = C_e APORm\ \Delta X\ \Delta Y 62.4/1E6 \tag{54}$$

$$PL_t = C_t APORm\ \Delta X\ \Delta Y 62.4/1E6 \tag{55}$$

For aquifer-aquitard system (see Prickett et al., 1981, pp. 51,52)

$$PL_s = C_s\ \Delta t(P'/m')\ \Delta X\ \Delta Y(h_{a1} - h_{a2})8.3453/1E6 \tag{56}$$

For aquifer layer system

$$PL_s = C_s\ \Delta t 16.6906\ \Delta X\ \Delta Y(h_{L1} - h_{L2})/[(m_1/P_{v1} + m_2/P_{v2})1E6] \tag{57}$$

where

$$C_s = (C_b + C_e)/2 \tag{58}$$

PL_a = total contaminant load in aquifer or aquifer layer at end of simulation time increment with areal leakage in pounds

PL_t = total contaminant load in aquifer or aquifer layer at end of simulation time increment without areal leakage in pounds

PL_s = total contaminant load leaving aquifer or aquifer layer during simulation time increment as the result of areal leakage in pounds

C_e = concentration in aquifer or aquifer layer at end of simulation time increment with areal leakage in mg/L

C_t = concentration in aquifer or aquifer layer at end of simulation time increment without areal leakage in mg/L

C_s = average concentration in aquifer or aquifer layer during simulation time increment with areal leakage in mg/L

C_b = concentration in aquifer or aquifer layer at start of simulation time increment in mg/L

Δt = simulation time increment in days

P' = aquitard vertical hydraulic conductivity in gpd/sq ft

m' = aquitard thickness in ft

h_{a1} = aquifer head elevation at node in ft

h_{a2} = sourcebed head elevation at node in ft

ΔX = grid spacing in X-direction in ft

ΔY = grid spacing in Y-direction in ft

P_{V1} = aquifer layer 1 vertical hydraulic conductivity in gpd/sq ft

P_{V2} = aquifer layer 2 vertical hydraulic conductivity in gpd/sq ft

m_1 = aquifer layer 1 thickness in ft

m_2 = aquifer layer 2 thickness in ft

h_{L1} = aquifer layer 1 head elevation at node in ft

h_{L2} = aquifer layer 2 head elevation at node in ft

APOR = actual aquifer or aquifer layer porosity as a decimal

Equation 58 assumes that the average concentration in the aquifer or aquifer layer during the simulation time increment is the sum of the concentrations at the start and end of the simulation time increment divided by 2. Concentration at the start of the simulation time increment is affected by pre-existing contaminant conditions and any slug contaminant sources. Sink contaminant load is considered in Equation 53 by excluding particles captured by local sinks such as production wells and streams from computations of PL_t (see Prickett et al., 1981, p. 50). The simulation time increment and particle mass should be as small as possible so that values of C_s are reasonably precise. Aquifer or aquifer layer heads mentioned above are average heads during the simulation time increment.

Substituting Equations 54–58 into Equation 53 and solving that equation for C_e results in the following expression:

$$C_e = (2C_t - C_bF1)/(F1 + 2) \qquad (59)$$

where

For aquifer-aquitard system

$$F1 = \Delta t(P'/m')(h_{a1} - h_{a2})\ 8.3453/(APORm62.4) \qquad (60)$$

For aquifer layer system

$$F1=\Delta t(h_{L1}-h_{L2})16.6906/[(m_1/P_{v1}+m_2/P_{v2})APORm_1 62.4] \qquad (61)$$

Knowing C_e, the total nodal areal sink contaminant load to be exported to subroutine AREA3D as an areal source load may be calculated with Equation 56 or 57 and the contaminant load left in the aquifer or aquifer layer with areal leakage may be calculated with Equation 54.

The average contaminant travel time (thickness divided by velocity) with or without adsorption through the aquitard or between aquifer layers, assuming vertical dispersion is negligible, may be calculated with the following equation (see Prickett et al., 1981, p. 7):

For aquifer-aquitard system

$$t_t = 7.48 \text{ EPOR } R_{d1v}m'^2/[P'(h_{a1} - h_{a2})] \tag{62}$$

For aquifer layer system

$$t_t = 7.48 \text{ EPOR } R_{d1}m_{av}^2/[P_{va}(h_{L1} - h_{L2})] \tag{63}$$

where

$$P_{va}=(m_1+m_2)/[(m_1/P_{v1})+(m_2/P_{v2})] \tag{64}$$

(see Raudkivi and Callander, 1976, p. 118)

t_t = average contaminant travel time through aquitard in days

EPOR = aquifer or aquifer layer effective porosity as a decimal

R_{d1} = aquifer or aquifer layer retardation factor as a decimal

R_{d1v} = aquitard retardation factor as a decimal

$m_{av} = (m_1+m_2)/2$

It is assumed that fairly uniform vertical head losses occur within the chosen grid area. Vertical head losses must reflect average head conditions during the travel time. Heads in the aquifer or aquifer layer receiving contaminant load from areal sinks should reflect average head conditions during the simulation time increment after the travel time specified in subroutine ARESNK. The simulation time increment for the areal source should be the same as that used in ARESNK.

The data base for subroutine ARESNK is user-defined in subroutine CONSTR and the MAIN program. The user may create an ASCII file or hard copy of nodal areal sink loads.

Random Number Generator

A random number generator is not included in Microsoft's list of intrinsic functions, but is included in GWTR3D as function RAN. This function returns a real random number equal to or greater than zero and less than one. RAN is based on ideas in the program presented by Wagener (1980, p. 176) and the linear congruential method involving three integer constants described by Knuth (1969). RAN seems to work well with the IBM PC; other random number generators may work better on other microcomputers. Some random generators in the literature are inappropriate for GWTR3D.

Appendix A
Source Code for GWFL3D

The source code, created with the Line Editor Edlin included in the MS-Disk Operating System (trademark of Microsoft Corp.), for GWFL3D is listed in this appendix. Instructions for using Edlin to create and edit files are provided in the MS-Disk Operating System manual and by Wolverton (1984, pp. 196–231). The program is written in FORTRAN 77 (see Etter, 1987 and Fouri et al., 1986) specifically for the Microsoft FORTRAN Compiler version 3.2 (trademark of Microsoft Corp.) and an IBM PC (trademark of International Business Machines Corp.). Program algorithms are documented as COMMENT statements in lists to assist the reader in understanding and modifying the codes. Many statements in the programs could be restructured or eliminated to produce more efficient coding, but they have been intentionally written in their present form to provide clear explanations of processing steps. The finite-difference grid is limited to 20x20 because of limited microcomputer core storage. The grid may be enlarged by redimensioning appropriate arrays if more core storage is available. GWFL3D consists of a main program and 17 subroutines filling 141273 bytes of memory. The number of routines was made large in an effort to make the program understandable to inexperienced users and to enable the use of EDLIN which is designed for short files. The experienced user may wish to consolidate some of the subroutines with a more sophisticated word processor.

GWFL3D SUBROUTINES

DATPR
DRAIN
EVAP
FILEI
FILE0
FWEL
INFO1
INFO2
MINE
PARPEN
PROFIL
RESLT
RIVER
SIMEQ
WATBL
WELSTOR
XYZDT

```
      PROGRAM GWFL3D
C
C     METACOMMANDS
C
$LARGE
C
C     DECLARATIONS
C
      CHARACTER*14FNAME
      DIMENSION H(20,20,4),HO(20,20,4),SF1(20,20,4),Q(20,20,4),
     *T1(20,20,4),T2(20,20,4),P1(20,20,4),P2(20,20,4),DL(20,20,4),
     *THICK(20,20,4),THICKC(20,20,4),P3(20,20,4),R(20,20,4),B(21,4),
     *G(21,4),HH(4),PP(4),PM(4),PMC(4),DELX(21),DELY(21),RR(4),SS1(4),
     *PPC(4),TIME(50),DELTA(50),S1(20,20,4),P4(20,20,4),PPA(4),
     *IP(50),JP(50),KP(50),IO(25),JO(25),KO(25),HTD(24,50),S3(20,20,4),
     *STORE(4),PUMP(4),DIFF(4),ALU(4),ALD(4),TOTIN(4),TOTOUT(4),
     *STORE1(4),RS(20,20,4),SS3(4),SS2(4),S2(20,20,4),SF2(20,20,4),
     *HD(20,20,20),ALPHA(20,20,1),QDY(20,20,1),A2(4),INODE(20),
     *RH(20,20,1),RD(20,20,1),QP(20,20,4),QI(20,20,4),QR(20,20,1),
     *IR(50),JR(50),KR(50),MFW(20,20,4),MIF(50),MJF(50),MKF(50),
     *CELE(20,20,4),QF(20,20,4),FWR(50),FWC(50),QMD(20,20,4),JNODE(20),
     *BOT(20,20,4),MSB(20,20,1),MDR(20,20,1),QE(20,20,1),RINJ(4),
     *RRECH(1),FWELL(4),DMINE(4),SRECH(1),DPUMP(1),
     *EVAPO(1),QWELL(4),
     *CH(20,20,4),QRD(20,20,4),QST(20,20,4),QDRAIN(20,20,1),XSUM(20),
     *YSUM(20),CHH(4),BOTT(4),MQE(20,20,1),RMA(4),A1(4),HHE(20,20,4),
     *HNODE(50,4)
      CALL INFO1
C
C     INTERACTIVE DATA BASE INPUT
C
      WRITE(*,20)
   20 FORMAT('0','MODEL DATA BASE SPECIFICATION:'/)
      WRITE(*,21)
   21 FORMAT(1X,'Please be careful in responding to interactive'/1X,
     *'data base prompts. Option prompts will be repeated if you'/1X,
     *'respond incorrectly. Be sure you enter either a real or'/1X,
     *'integer number as specified in prompt statement. TURN ON'/1X,
     *'THE PRINTER OR ELSE AN ERROR WILL OCCUR.'/)
   22 WRITE(*,23)
   23 FORMAT(1X,'Enter 1 for keyboard time constant data'/1X,
     *'basc input or 2 to import data base file previously'/1X,
     *'created with this program  '\)
      READ(*,*)KFI
      IF(KFI.NE.1.AND.KFI.NE.2) GOTO 22
      IF(KFI.EQ.2) THEN
```

```
      CALL FILEI(NC,NR,NSTEPS,NP,NOBS,ERROR,WBE,TIME,DELTA,IP,
     *JP,KP,IO,JO,KO,NL,H,THICK,P1,P2,T1,T2,P3,THICKC,SF1,
     *S1,R,DELX,DELY,Q,P4,S3,MAS,S2,RS,DL,MF1,MNF,MIF,
     *MJF,MKF,MFW,CELE,SF2,CH,BOT,NRW,IR,JR,KR,MUS,HSB,
     *MAL,MAU,KFI,MDY,LC1,QI,QR,QMD,MSB,MDR,QE,QP,
     *QF,QST,QDRAIN,RH,RD)
      OPEN(4,FILE='LPT1:')
      GOTO 392
      ENDIF
      WRITE(*,24)
   24 FORMAT('0','Supported English units are: gpd,gpm,cfs,days,'/1X,
     *'min,ft,sq ft,cu ft,gpd/sq ft,gpd/ft. Unit requirements are'/1X,
     *'cited in data base input statements. If your units do'/1X,
     *'not match these units then use conversion tables to obtain'/1X,
     *'specified units.')
C
C     DEFINE OUTPUT DEVICE
      OPEN(4,FILE='LPT1:')
C
   25 WRITE(*,26)
   26 FORMAT('0','Enter 1 for display of data base background'/1X,
     *'information at selected intervals in program or 2 for'/1X,
     *'no display  '\)
      READ(*,*)LC1
      IF(LC1.NE.1.AND.LC1.NE.2)GOTO 25
      WRITE(4,27)
   27 FORMAT('0','MODEL DATA BASE SPECIFICATION:'/)
C
C     DEFINE TIME INCREMENTS
C
      WRITE(*,30)
   30 FORMAT('0','Enter integer number of time increments,'/1X,
     *'six time increments must precede earliest time of interest'/1X,
     *'for precise head or drawdown calculation, must be < 51,'/1X,
     *'if delayed gravity drainage is simulated then must be <21  '\)
      READ(*,*)NSTEPS
      WRITE(4,32)NSTEPS
   32 FORMAT(1X,'Number of time increments=',I3)
      IF(LC1.EQ.2)GOTO 35
      WRITE(*,33)
   33 FORMAT('0',
     *'The finite-difference grid lines superposed over the'/1X,
     *'aquifer system map are indexed using the I (column),'/1X,
     *'J (row) notation colinear with the X and Y directions,'/1X,
     *'respectively. Coordinate axes intersect at the upper'/1X,
     *'left corner of the map. I coordinates increase left to'/1X,
     *'right and J coordinates increase top to bottom. Grid'/1X,
     *'spacing is centered at the nodes (grid line intersections).'/1X,
     *'In multi-layered aquifer systems, layers are numbered'/1X,
     *'from land surface downward using the K (layer) notation'/1X,
     *'colinear with the Z direction. The origin of the system'/1X,
     *'is the upper-left corner of the topmost layer. The'/1X,
```

```
     *'number of aquifers or aquifer layers must be <5.'/1X,
     *'Grid spacing is the same in all aquifer system'/1X,
     *'layers. Barrier boundaries occur along the borders of'/1X,
     *'the grid one half grid spacing beyond end grid lines.'/)
      PAUSE
      WRITE(*,34)
   34 FORMAT('0',
     *'Grid spacings and time increments are interdependent'/1X,
     *'and should be jointly specified so that head changes'/1X,
     *'at nodes adjacent to sink/source nodes are appreciable'/1X,
     *'during the initial time increment. Grid spacing should'/1X,
     *'increase gradually outward from sink/source nodes.'/1X,
     *'Small time increments should be specified each time'/1X,
     *'sink flow rates, source flow rates, or aquifer system'/1X,
     *'features are modified. Time increments should then be'/1X,
     *'increased as a gemetric progression of ratio 1.2 until'/1X,
     *'the next modification. The initial time increment is'/1X,
     *'selected based on the grid spacing, hydraulic character-'/1X,
     *'istics, boundary conditions, and the period of interest'/1X,
     *'taking into account that 6 time increments should precede'/1X,
     *'first time of interest for precice head calculations.')
   35 DO 45 I=1,NSTEPS
      WRITE(*,36)I
   36 FORMAT('0','Enter a real number value for the time at'/1X,
     *'the end of time increment',I3,' in days  '\)
      READ(*,*)TIME(I)
      IF(I.EQ.1)THEN
      DELTA(I)=TIME(I)
      GOTO 37
      ENDIF
      DELTA(I)=TIME(I)-TIME(I-1)
   37 WRITE(*,40)DELTA(I),I
   40 FORMAT(1X,'Time increment in days=',F10.3,/1X,
     *'for time increment number',I3)
      WRITE(4,41)I
   41 FORMAT(1X,'Time increment number=',I3)
      WRITE(4,42)TIME(I)
   42 FORMAT(1X,'Time at end of time increment in days=',F12.5)
      WRITE(4,43)DELTA(I)
   43 FORMAT(1X,'Time increment in days=',F12.5)
   45 CONTINUE
   46 WRITE(*,47)
   47 FORMAT('0','Enter 1 to revise increment time data base or'/1X,
     *'2 to continue with data base entry  '\)
      READ(*,*)NREV
      IF(NREV.NE.1.AND.NREV.NE.2) GOTO 46
      IF(NREV.EQ.1) GOTO 35
C
C     DEFINE MODEL GRID
C
      IF(KFI.EQ.2)THEN
   48 WRITE(*,49)
```

```
   49 FORMAT('0','Enter 1 to redefine model grid or 2'/1X,
     *'to continue program  '\)
      READ(*,*)LDM
      IF(LDM.NE.1.AND.LDM.NE.2)GOTO 48
      IF(LDM.EQ.2)GOTO 93
      ENDIF
   52 WRITE(*,54)
   54 FORMAT('0','Enter integer number of grid columns'/1X,
     *'must be <21  '\)
      READ(*,*)NC
      WRITE(4,55)NC
   55 FORMAT(1X,'Number of grid columns=',I3)
      WRITE(*,56)
   56 FORMAT(1X,'Enter integer number of grid rows'/1X,
     *'must be <21  '\)
      READ(*,*)NR
      WRITE(4,57)NR
   57 FORMAT(1X,'Number of grid rows=',I3)
      WRITE(*,58)
   58 FORMAT(1X,'The default grid setting is uniform (square).'/1X,
     *'Enter the real number default grid spacing in ft,'/1X,
     *'usually ranges from 10-10,000 ft  '\)
      READ(*,*)DGS
      DO 62 I=1,NC+1
      DELX(I)=DGS
   62 CONTINUE
      DO 64 J=1,NR+1
      DELY(J)=DGS
   64 CONTINUE
   65 WRITE(*,66)
   66 FORMAT('0','Enter 1 for a variable grid or 2 to retain'/1X,
     *'default grid setting. Please read prompt'/1X,
     *'statements carefully before responding  '\)
      READ(*,*)NVGS
      IF(NVGS.NE.1.AND.NVGS.NE.2) GOTO 65
      IF(NVGS.EQ.2) GOTO 93
      IF(LC1.EQ.2)GOTO 69
      WRITE(*,67)
   67 FORMAT('0','Grid size is partly governed by the desired'/1X,
     *'precision of numerical approximation; the smaller the'/1X,
     *'grid size the better the approximation. Grid spacings'/1X,
     *'along individual columns and rows are uniform. The'/1X,
     *'ratio of adjacent grid spacings is usually less than'/1X,
     *'5 except for high precision head or drawdown'/1X,
     *'calculations when it is 2 or less.')
C
C     DEFINE VARIABLE GRID SIZES
C
   69 WRITE(*,70)
   70 FORMAT('0','Enter integer number of column grid blocks'/1X,
     *'to be changed to variable grid settings  '\)
      READ(*,*)NGB
```

```
      DO 79 K=1,NGB
      WRITE(*,72)K
   72 FORMAT(1X,'Block number=',I3)
      WRITE(*,74)
   74 FORMAT(1X,'Enter integer grid block left- and right-'/1X,
     *'most column I coordinates separated by a comma  '\)
      READ(*,*)IVG1,IVG2
      DO 78 I=IVG1,IVG2
      WRITE(*,76)I
   76 FORMAT(1X,'Enter real number grid spacing in ft'/1X,
     *'starting at column',I3,'  '\)
      READ(*,*)DELX(I)
   78 CONTINUE
   79 CONTINUE
      WRITE(*,81)
   81 FORMAT('0','Enter integer number of row grid blocks'/1X,
     *'to be changed to variable grid settings  '\)
      READ(*,*)NGB
      DO 90 K=1,NGB
      WRITE(*,82)K
   82 FORMAT(1X,'Block number=',I3)
      WRITE(*,83)
   83 FORMAT(1X,'Enter integer grid block upper- and lower-'/1X,
     *'most row J coordinates separated by a comma  '\)
      READ(*,*)JVG1,JVG2
      DO 85 J=JVG1,JVG2
      WRITE(*,84)J
   84 FORMAT(1X,'Enter real number grid spacing in ft'/1X,
     *'starting at row',I3,'  '\)
      READ(*,*)DELY(J)
   85 CONTINUE
   90 CONTINUE
   91 WRITE(*,92)
   92 FORMAT('0','Enter 1 to revise variable grid data base or'/1X,
     *'2 to continue data base input  '\)
      READ(*,*)NRGB
      IF(NRGB.NE.1.AND.NRGB.NE.2) GOTO 91
      IF(NRGB.EQ.1) GOTO 69
   93 CALL INFO2(MAL,MAU,MAS,MDY,NL,LC1)
   94 WRITE(*,96)
   96 FORMAT('0','Enter 1 to simulate leaky artesian conditions'/1X,
     *'in uppermost unit or 2 to simulate nonleaky or water table'/1X,
     *'conditions in uppermost unit.  '\)
      READ(*,*)MUS
      IF(MUS.NE.1.AND.MUS.NE.2)GOTO 94
      IF(MUS.EQ.1)THEN
      WRITE(*,98)
   98 FORMAT('0','Enter real number constant head in sourcebed'/1X,
     *'above uppermost unit in ft  '\)
      READ(*,*)HSB
      WRITE(4,100)HSB
  100 FORMAT(1X,'Constant head in sourcebed above'/1X,
```

```
     *'uppermost unit in ft=',F9.2)
      ENDIF
      IF(LC1.EQ.2)GOTO 190
      WRITE(*,120)
  120 FORMAT('0','Grid arrays will first be filled with default'/1X,
     *'hydraulic characteristic values to avoid the inconven-'/1X,
     *'ience of individually specifying node values when they'/1X,
     *'are the same for large portions of the grid. Data is'/1X,
     *'thereby provided for simulating an infinite aquifer system'/1X,
     *'having homogeneous-isotropic conditions. The opportun-'/1X,
     *'ity to simulate an aquifer system having variable'/1X,
     *'hydraulic characteristic values and boundaries will'/1X,
     *'then be provided by possible replacement of any default'/1X,
     *'value on a node-by-node basis.'/)
      PAUSE
C
C     FILL ARRAYS WITH DEFAULT HYDRAULIC CHARACTERISTIC
C     VALUES
C
  190 IF(KFI.EQ.2)THEN
  192 WRITE(*,194)
  194 FORMAT('0','Enter 1 to redefine default hydraulic'/1X,
     *'characteristics values or 2 to continue program  '\)
      READ(*,*)LDH
      IF(LDH.NE.1.AND.LDH.NE.2)GOTO 192
      IF (LDH.EQ.2)GOTO 243
      ENDIF
  197 WRITE(*,198)
  198 FORMAT('0','After each of the following aquifer system'/1X,
     *'hydraulic characteristics and geometries is displayed,'/1X,
     *'enter the real number default value. The process is'/1X,
     *'repeated for each aquifer or aquifer layer. Elevations'/1X,
     *'refer to mean sea level or zero base level. If a zero'/1X,
     *'base level is used, some elevations will be negative'/1X,
     *'and drawdown or recovery instead of head will be'/1X,
     *'calculated and displayed.')
      DO 242 K=1,NL
      WRITE(*,199)K
  199 FORMAT('0','Unit or layer number=',I2)
      WRITE(*,200)
  200 FORMAT(1X,'Aquifer horizontal hydraulic conductivity'/1X,
     *'in gpd/sq ft, usually ranges from 1 to'/1X,
     *'10000 gpd/sq ft  '\)
      READ(*,*)PP(K)
      IF(MDY.EQ.1)GOTO 201
      IF(MAL.EQ.2) THEN
      PPA(K)=0.0
      GOTO 203
      ENDIF
      IF(MAU.EQ.1.AND.K.GT.2)THEN
      PPA(K)=0.0
      GOTO 203
```

```
      ENDIF
      IF(MAU.EQ.2.AND.K.LT.3)THEN
      PPA(K)=0.0
      GOTO 203
      ENDIF
201 WRITE(*,202)
202 FORMAT(1X,'Aquifer vertical hydraulic conductivity'/1X,
   *'in gpd/sq ft, usually ranges from 1/10 to 1/100'/1X,
   *'of aquifer horizontal hydraulic conductivity  '\)
      READ(*,*)PPA(K)
203 WRITE(*,204)
204 FORMAT(1X,'Aquifer artesian storativity as a fraction,'/1X,
   *'usually ranges from 1E-6 to 1E-3  '\)
      READ(*,*)SS1(K)
      WRITE(*,206)
206 FORMAT(1X,'Aquifer water table specific yield as a decimal,'/1X,
   *'usually ranges from .005 to .4  '\)
      READ(*,*)SS2(K)
      WRITE(*,208)
208 FORMAT(1X,'Aquifer thickness in ft  '\)
      READ(*,*)PM(K)
      WRITE(*,209)
209 FORMAT(1X,'Aquifer top elevation in ft  '\)
      READ(*,*)CHH(K)
      WRITE(*,210)
210 FORMAT(1X,'Aquifer base elevation in ft  '\)
      READ(*,*)BOTT(K)
      WRITE(*,214)
214 FORMAT(1X,'Elevation of initial head in aquifer'/1X,
   *'in ft  '\)
      READ(*,*)HH(K)
      IF(MAU.EQ.3)THEN
      PPC(K)=0.0
      GOTO 216
      ENDIF
      IF(MAU.EQ.1.AND.K.LT.3)THEN
      PPC(K)=0.0
      GOTO 216
      ENDIF
      IF(MAU.EQ.2.AND.K.GT.2)THEN
      PPC(K)=0.0
      GOTO 216
      ENDIF
      IF(MUS.EQ.2.AND.K.EQ.1)THEN
      PPC(K)=0.0
      GOTO 216
      ENDIF
      WRITE(*,215)
215 FORMAT(1X,'Aquitard vertical hydraulic conductivity in'/1X,
   *'gpd/sq ft, usually ranges from 1E-7 to 10'/1X,
   *'gpd/sq ft  '\)
      READ(*,*)PPC(K)
```

```
216 IF(MAS.EQ.1)GOTO 218
    IF(MAS.EQ.2)THEN
    SS3(K)=0.0
    GOTO 223
    ENDIF
    IF(MAU.EQ.3)THEN
    SS3(K)=0.0
    GOTO 223
    ENDIF
    IF(MAU.EQ.1.AND.K.LT.3)THEN
    SS3(K)=0.0
    GOTO 223
    ENDIF
    IF(MAU.EQ.2.AND.K.GT.2)THEN
    SS3(K)=0.0
    GOTO 223
    ENDIF
218 IF(MUS.EQ.2.AND.K.EQ.1)THEN
    SS3(K)=0.0
    GOTO 223
    ENDIF
    WRITE(*,221)
221 FORMAT(1X,'Aquitard storativity as a decimal,'/1X,
   *'usually ranges from 1E-6 to 1E-3  '\)
    READ(*,*)SS3(K)
223 IF(MAU.EQ.3)THEN
    PMC(K)=0.0
    GOTO 230
    ENDIF
    IF(MAU.EQ.1.AND.K.LT.3)THEN
    PMC(K)=0.0
    GOTO 230
    ENDIF
    IF(MAU.EQ.2.AND.K.GT.2)THEN
    PMC(K)=0.0
    GOTO 230
    ENDIF
    IF(MUS.EQ.2.AND.K.EQ.1)THEN
    PMC(K)=0.0
    GOTO 230
    ENDIF
    WRITE(*,224)
224 FORMAT(1X,'Aquitard thickness in ft  '\)
    READ(*,*)PMC(K)
230 IF(K.EQ.NL)THEN
236 WRITE(*,237)
237 FORMAT('0','Enter 1 to revise default values or'/1X,
   *'2 to continue with data base specification  '\)
    READ(*,*)NRDV
    IF (NRDV.NE.1.AND.NRDV.NE.2) GO TO 236
    IF(NRDV.EQ.1)GOTO 197
    ENDIF
```

```
      DO 240 J=1,NR
      DO 238 I=1,NC
      P1(I,J,K)=PP(K)
      P2(I,J,K)=PP(K)
      P3(I,J,K)=PPC(K)
      P4(I,J,K)=PPA(K)
      S1(I,J,K)=SS1(K)
      S2(I,J,K)=SS2(K)
      S3(I,J,K)=SS3(K)
      THICK(I,J,K)=PM(K)
      H(I,J,K)=HH(K)
      THICKC(I,J,K)=PMC(K)
      T1(I,J,K)=P1(I,J,K)*THICK(I,J,K)
      T2(I,J,K)=P2(I,J,K)*THICK(I,J,K)
      CH(I,J,K)=CHH(K)
      BOT(I,J,K)=BOTT(K)
C
C     INITIALIZE VARIABLES
C
      R(I,J,K)=0.0
      Q(I,J,K)=0.0
      DL(I,J,K)=0.0
      RS(I,J,K)=0.0
      QI(I,J,K)=0.0
      IF(K.EQ.1)QR(I,J,K)=0.0
      QMD(I,J,K)=0.0
      MFW(I,J,K)=0
      CELE(I,J,K)=0.0
      IF(K.EQ.1)MSB(I,J,1)=0
      IF(K.EQ.1)MDR(I,J,1)=0
      IF(K.EQ.1)QE(I,J,K)=0.0
      QP(I,J,K)=0.0
      QF(I,J,K)=0.0
      IF(K.EQ.1)QST(I,J,1)=0.0
      IF(K.EQ.1)QDRAIN(I,J,1)=0.0
      IF(K.EQ.1)RH(I,J,1)=0.0
      IF(K.EQ.1)RD(I,J,1)=0.0
  238 CONTINUE
  240 CONTINUE
  242 CONTINUE
  243 WRITE(*,244)
  244 FORMAT('0','Enter 1 to change default values at'/1X,
     *'selected nodes or 2 to retain default values. Please'/1X,
     *'read prompt statements carefully before responding  '\)
      READ(*,*)NCDV
      IF(NCDV.NE.1.AND.NCDV.NE.2) GOTO 243
      IF(NCDV.EQ.2) GOTO 297
      IF(LC1.EQ.2)GOTO 247
  245 WRITE(*,246)
  246 FORMAT('0','Conductivity is defined for a particular'/1X,
     *'prism of the aquifer system and for a particular'/1X,
     *'direction. Therefore, conductivity in the three'/1X,
```

```
     *'principal directions (X,Y,Z) may be different at'/1X,
     *'any given node. Some default values may need to be changed'/1X,
     *'as the simulation time period changes. These default'/1X,
     *'values will be addressed later in the program. After'/1X,
     *'each of the following aquifer system hydraulic char-'/1X,
     *'acteristics and geometries is displayed, enter the'/1X,
     *'real number nodal value. This process is repeated for'/1X,
     *'each aquifer-aquitard unit or aquifer layer.')
      PAUSE
C
C     DEFINE NODAL HYDRAULIC CHARACTERISTIC VALUES
C
  247 WRITE(*,248)
  248 FORMAT('0','Enter integer number of grid blocks'/1X,
     *'containing nodes whose default values are to be'/1X,
     *'changed  '\)
      READ(*,*)NGB
      DO 294 L=1,NGB
      WRITE(*,249)L
  249 FORMAT(1X,'Block number=',I3)
      WRITE(*,250)
  250 FORMAT(1X,'Enter integer grid block left- and right-'/1X,
     *'most column I coordinates separated by a comma  '\)
      READ(*,*)ICD1,ICD2
      WRITE(*,252)
  252 FORMAT(1X,'Enter integer grid block upper- and lower-'/1X,
     *'most row J coordinates separated by a comma  '\)
      READ(*,*)JCD1,JCD2
      WRITE(*,253)
  253 FORMAT(1X,'Enter integer number of units or layers'/1X,
     *'in which default values are to be changed (must'/1X,
     *'be <5)  '\)
      READ(*,*)NLC
      DO 292 K=1,NLC
      DO 290 I=ICD1,ICD2
      DO 289 J=JCD1,JCD2
      WRITE(*,254)K
  254 FORMAT('0','Aquifer unit or layer number',I3)
      WRITE(*,255)I
  255 FORMAT(1X,'I-coordinate of node',I3)
      WRITE(*,256)J
  256 FORMAT(1X,'J-coordinate of node',I3)
      WRITE(*,258)
  258 FORMAT(1X,'Aquifer horizontal conductivity'/1X,
     *'in I-direction, vector volume is unit prism centered'/1X,
     *'along I-grid line between I,J and I,J+1, in gpd'/1X,
     *'/sq ft  '\)
      READ(*,*)P1(I,J,K)
      WRITE(*,260)
  260 FORMAT(1X,'Aquifer horizontal conductivity'/1X,
     *'in J-direction, vector volume is unit prism centered'/1X,
     *'along J-grid line between I,J and I+1,J, in gpd'/1X,
```

```
   *'sq ft  '\)
    READ(*,*)P2(I,J,K)
    IF(MDY.EQ.1)GOTO 261
    IF(MAL.EQ.2)GOTO 263
    IF(MAU.EQ.1.AND.K.GT.2)GOTO 263
    IF(MAU.EQ.2.AND.K.LT.3)GOTO 263
261 WRITE(*,262)
262 FORMAT(1X,'Aquifer vertical hydraulic conductivity'/1X,
   *'in gpd/sq ft  '\)
    READ(*,*)P4(I,J,K)
263 WRITE(*,264)
264 FORMAT(1X,'Aquifer artesian storativity at'/1X,
   *'node, vector volume is prism with aquifer thickness'/1X,
   *'centered at node between points midway to adjacent'/1X,
   *'nodes, as a decimal  '\)
    READ(*,*)S1(I,J,K)
    WRITE(*,266)
266 FORMAT(1X,'Aquifer water table specific yield at'/1X,
   *'node, vector volume is prism with aquifer thickness'/1X,
   *'centered at node between points midway to adjacent'/1X,
   *'nodes, as a decimal  '\)
    READ(*,*)S2(I,J,K)
    WRITE(*,268)
268 FORMAT(1X,'Aquifer thickness in ft  '\)
    READ(*,*)THICK(I,J,K)
    WRITE(*,269)
269 FORMAT(1X,'Aquifer top elevation in ft  '\)
    READ(*,*)CH(I,J,K)
    WRITE(*,270)
270 FORMAT(1X,'Aquifer base elevation in ft  '\)
    READ(*,*)BOT(I,J,K)
    T1(I,J,K)=P1(I,J,K)*THICK(I,J,K)
    T2(I,J,K)=P2(I,J,K)*THICK(I,J,K)
    WRITE(*,274)
274 FORMAT(1X,'Elevation of initial head in'/1X,
   *'aquifer in ft  '\)
    READ(*,*)H(I,J,K)
    IF(MAU.EQ.3)GOTO 285
    IF(MAU.EQ.1.AND.K.LT.3)GOTO 285
    IF(MAU.EQ.2.AND.K.GT.2)GOTO 285
    IF(MUS.EQ.2.AND.K.EQ.1)GOTO 289
    WRITE(*,280)
280 FORMAT(1X,'Aquitard vertical hydraulic conductivity of'/1X,
   *'in gpd/ft  '\)
    READ(*,*)P3(I,J,K)
    WRITE(*,284)
284 FORMAT(1X,'Aquitard thickness in ft  '\)
    READ(*,*)THICKC(I,J,K)
285 IF(MAS.EQ.2)GOTO 289
    IF(MAU.EQ.3)GOTO 289
    IF(MAU.EQ.1.AND.K.LT.3)GOTO 289
    IF(MAU.EQ.2.AND.K.GT.2)GOTO 289
```

```
      WRITE(*,286)
  286 FORMAT(1X,'Aquitard storativity as a decimal  '\)
      READ(*,*)S3(I,J,K)
  289 CONTINUE
  290 CONTINUE
  292 CONTINUE
  294 CONTINUE
  295 WRITE(*,296)
  296 FORMAT('0','Enter 1 to revise default value'/1X,
     *'changes or 2 to continue data base specification  '\)
      READ(*,*)NREVC
      IF(NREVC.NE.1.AND.NREVC.NE.2) GOTO 295
      IF(NREVC.EQ.1) GOTO 245
C
C     DEFINE SINK/SOURCE GEOMETRIES
C
  297 WRITE(*,298)
  298 FORMAT('0','Some sink/source geometries may vary'/1X,
     *'with time and will be addressed later in this program'/1X,
     *'Wells are assumed to fully penetrate aquifers.'/)
  299 WRITE(*,300)
  300 FORMAT(1X,'Enter integer number of production wells'/1X,
     *'(must be <51)  '\)
      READ(*,*)NP
      WRITE(4,301)NP
  301 FORMAT(1X,'Number of production wells=',I3)
      DO 309 I=1,NP
      WRITE(*,302)I
  302 FORMAT(1X,'Production well number',I3)
      WRITE(*,304)
  304 FORMAT(1X,'Enter integer number of unit or'/1X,
     *'layer in which the production well is located  '\)
      READ(*,*)KP(I)
      WRITE(*,306)
  306 FORMAT(1X,'Enter integer I-coordinate of production'/1X,
     *'well  '\)
      READ(*,*)IP(I)
      WRITE(*,308)
  308 FORMAT(1X,'Enter integer J-coordinate of production'/1X,
     *'well  '\)
      READ(*,*)JP(I)
  309 CONTINUE
      WRITE(*,310)
  310 FORMAT('0','The head calculated for the production well node'/1X,
     *'is the average head for the well grid block and'/1X,
     *'not the head in the production well.')
      WRITE(*,319)
  319 FORMAT('0','Enter integer number of injection wells'/1X,
     *'(must be <51)  '\)
      READ(*,*)NRW
      WRITE(4,320)NRW
  320 FORMAT(1X,'Number of injection wells=',I3)
```

```
      IF(NRW.EQ.0)GOTO 328
      WRITE(*,321)
  321 FORMAT('0','The head calculated for the injection well'/1X,
     *'node is the average head for the well grid block and'/1X,
     *'not the head in the injection well.')
      DO 326 I=1,NRW
      WRITE(*,322)I
  322 FORMAT('0','Injection well number=',I3)
      WRITE(*,323)
  323 FORMAT(1X,'Enter integer number I-coordinate of'/1X,
     *'injection well  '\)
      READ(*,*)IR(I)
      WRITE(*,324)
  324 FORMAT(1X,'Enter integer number J-coordinate of'/1X,
     *'injection well  '\)
      READ(*,*)JR(I)
      WRITE(*,325)
  325 FORMAT(1X,'Enter integer unit or layer number in'/1X,
     *'which injection well is open  '\)
      READ(*,*)KR(I)
  326 CONTINUE
  328 CALL FWEL(MNF,MFW,MIF,MJF,CELE,FWR,FWC,T1,T2,DELX,DELY,
     *R,MF1,MKF)
C
C     DEFINE TIME-HEAD TABLE OBSERVATION WELLS
C
  332 WRITE(*,334)
  334 FORMAT('0','Enter integer number of observation wells'/1X,
     *'for which time-head tables are to be prepared'/1X,
     *'(must be <25)  '\)
      READ(*,*)NOBS
      WRITE(4,335)NOBS
  335 FORMAT(1X,'Number of time-head observation wells=',I3)
      DO 342 I=1,NOBS
      WRITE(*,336)I
  336 FORMAT(1X,'Observation well number',I3)
      WRITE(*,337)
  337 FORMAT(1X,'Enter integer number of unit or layer'/1X,
     *'in which the observation well is located  '\)
      READ(*,*)KO(I)
      WRITE(*,338)
  338 FORMAT(1X,'Enter integer I-coordinate of observation'/1X,
     *'well  '\)
      READ(*,*)IO(I)
      WRITE(*,340)
  340 FORMAT(1X,'Enter integer J-coordinate of observation'/1X,
     *'well  '\)
      READ(*,*)JO(I)
  342 CONTINUE
  344 WRITE(*,346)
  346 FORMAT('0','Enter 1 to revise sink/source geometry'/1X,
     *'and observation well definition data base or 2 to'/1X,
```

```
     *'continue data base specification  '\)
      READ(*,*)NSS
      IF(NSS.NE.1.AND. NSS.NE.2)GOTO 344
      IF(NSS.EQ.1)GOTO 297
C
C     DEFINE SIMULATION CONDITIONS
C
      IF(MUS.EQ.2)HSB=0.0
  351 WRITE(*,352)
  352 FORMAT('0','Enter 1 to simulate water table conditions'/1X,
     *'without delayed gravity yield in uppermost aquifer unit'/1X,
     *'or layer at the top of the aquifer system or 2 to simulate'/1X,
     *'water table conditions with delayed gravity yield or'/1X,
     *'artesian conditions  '\)
      READ(*,*)M11
      IF(M11.NE.1.AND.M11.NE.2)GOTO 351
      IF(M11.EQ.1)THEN
      DO 355 J=1,NR
      DO 354 I=1,NC
      S1(I,J,1)=S2(I,J,1)
  354 CONTINUE
  355 CONTINUE
      ENDIF
C
C     DEFINE CONVERGENCE TOLERANCES
C
      WRITE(*,362)
  362 FORMAT('0','Enter real number nodal head convergence error'/1X,
     *'tolerance in ft, usually .01 ft or 0.001 ft with delayed'/1X,
     *'gravity yield simulation.  '\)
      READ(*,*)HCE
      ERROR=HCE*NC*NR*NL
      WRITE(*,363)
  363 FORMAT(1X,'Enter water balance convergence error'/1X,
     *'tolerance in percent,usually 5%  '\)
      READ(*,*)WBE
C
C     ADJUST HYDRAULIC CHARACTERISTICS FOR VARIABLE GRID
C
      DO 386 K=1,NL
      DO 385 I=1,NC
      DO 384 J=1,NR
      P1(I,J,K)=P1(I,J,K)*2*DELX(I)/(DELY(J)+DELY(J+1))
      P2(I,J,K)=P2(I,J,K)*2*DELY(J)/(DELX(I)+DELX(I+1))
      T1(I,J,K)=P1(I,J,K)*THICK(I,J,K)
      T2(I,J,K)=P2(I,J,K)*THICK(I,J,K)
      SF1(I,J,K)=S1(I,J,K)*7.48*DELX(I)*DELY(J)
      SF2(I,J,K)=S2(I,J,K)*7.48*DELX(I)*DELY(J)
      IF(MAU.EQ.1.AND.K.EQ.2)GOTO 365
      IF(MAU.EQ.2.AND.K.EQ.4)GOTO 365
      IF(MAU.EQ.3.AND.K.GT.1)GOTO 365
      IF(MFW(I,J,K).EQ.1)GOTO 384
```

```
      IF(P3(I,J,K).EQ.0.0.OR.THICKC(I,J,K).EQ.0.0)THEN
      R(I,J,K)=0.0
      GOTO 384
      ENDIF
      R(I,J,K)=(P3(I,J,K)/THICKC(I,J,K))*DELX(I)*DELY(J)
      GOTO 384
365   R(I,J,K)=2*P4(I,J,K-1)*P4(I,J,K)*DELX(I)*DELY(J)/
     *(THICK(I,J,K-1)*P4(I,J,K)+THICK(I,J,K)*P4(I,J,K-1))
384   CONTINUE
385   CONTINUE
386   CONTINUE
390   WRITE(*,391)
391   FORMAT('0','Enter 1 to store data base on a diskette'/1X,
     *'as an ASCII file to be exported for later use with'/1X,
     *'this program or 2 to continue program  '\)
      READ(*,*)NST
      IF(NST.NE.1.AND.NST.NE.2)GOTO 390
      IF(NST.EQ.1)THEN
      CALL FILEO(NC,NR,NSTEPS,NP,NOBS,ERROR,WBE,TIME,DELTA,
     *IP,JP,KP,IO,JO,KO,NL,H,THICK,P1,P2,T1,T2,P3,THICKC,
     *SF1,S1,R,DELX,DELY,Q,P4,S3,MAS,S2,RS,DL,MF1,MNF,
     *MIF,MJF,MKF,MFW,CELE,SF2,CH,BOT,NRW,IR,JR,KR,MUS,HSB,
     *MAL,MAU,KFI,MDY,LC1,QI,QR,QMD,MSB,MDR,QE,QP,
     *QF,QST,QDRAIN,RH,RD)
      ENDIF
392   IF(KFI.EQ.2)THEN
393   WRITE(*,394)
394   FORMAT('0','Enter 1 to revise imported data base'/1X,
     *'or 2 to continue program  '\)
      READ(*,*)LDB
      IF(LDB.NE.1.AND.LDB.NE.2)GOTO 393
      IF (LDB.EQ.1)GOTO 25
      ENDIF
395   WRITE(*,396)
396   FORMAT('0','Enter 1 for display of intermediate'/1X,
     *'computation results or 2 for no display  '\)
      READ(*,*)NIC
      IF(NIC.NE.1.AND.NIC.NE.2)GOTO 395
      DO 399 K=1,NL
      DO 398 J=1,NR
      DO 397 I=1,NC
      HO(I,J,K)=0.0
397   CONTINUE
398   CONTINUE
399   CONTINUE
400   WRITE(*,401)
401   FORMAT('0','Enter 1 for hard copy of grid spacings'/1X,
     *'or 2 for no hard copy of grid spacings  '\)
      READ(*,*)MH1
      IF(MH1.NE.1.AND.MH1.NE.2)GOTO 400
      IF(MH1.EQ.2)GOTO 417
      WRITE(4,406)
```

```
  406 FORMAT('0',5X,'Column Number',3X,'Grid Spacing (ft)')
      DO 410 I=1,NC
      WRITE(4,408)I,DELX(I)
  408 FORMAT(1X,8X,I3,13X,F9.2)
  410 CONTINUE
      WRITE(4,412)
  412 FORMAT('0',5X,'Row Number',6X,'Grid Spacing (ft)')
      DO 416 J=1,NR
      WRITE(4,414)J,DELY(J)
  414 FORMAT(1X,8X,I3,13X,F9.2)
  416 CONTINUE
  417 DO 420 J=1,NR
      DO 418 I=1,NC
      IF(P4(I,J,1).EQ.0.0)THEN
      ALPHA(I,J,1)=0.0
      GOTO 418
      ENDIF
      ALPHA(I,J,1)=3*P4(I,J,1)/(S2(I,J,1)*THICK(I,J,1)*7.48)
  418 CONTINUE
  420 CONTINUE
      DO 424 K=1,NL
      DO 422 N=1,21
      G(N,K)=0.0
      B(N,K)=0.0
  422 CONTINUE
  424 CONTINUE
C
C     HARD COPY OF NODAL DATA BASE
C
  436 WRITE(*,438)
  438 FORMAT('0','Enter 1 for a hard copy of nodal hydraulic'/1X,
     *'characteristics and sink/source geometries or 2'/1X,
     *'to continue program   '\)
      READ(*,*)NHC
      IF(NHC.NE.1.AND.NHC.NE.2)GOTO 436
      IF(NHC.EQ.2)GOTO 460
      DO 459 K=1,NL
      WRITE(4,439)K
  439 FORMAT('0','Aquifer unit or layer number',I2)
      WRITE(4,440)
  440 FORMAT('0','Hydraulic Conductivity in I-Direction'/1X,
     *'in gpd/sq ft:')
      NZ=1
      CALL DATPR(NC,NR,NL,P1,K,NZ)
      WRITE(4,441)
  441 FORMAT('0','Hydraulic Conductivity in J-Direction'/1X,
     *'in gpd/sq ft:')
      CALL DATPR(NC,NR,NL,P2,K,NZ)
      IF(MAL.EQ.1)THEN
      IF(MAU.EQ.2.AND.K.LT.2)GOTO 443
      IF(MAU.EQ.1.AND.K.GT.2)GOTO 443
      WRITE(4,442)
```

```
442 FORMAT('0','Vertical hydraulic conductivity in gpd/sq ft:')
    CALL DATPR(NC,NR,NL,P4,K,NZ)
443 ENDIF
    WRITE(4,444)
444 FORMAT('0','Aquifer Thickness in ft:')
    CALL DATPR(NC,NR,NL,THICK,K,NZ)
    WRITE(4,445)
445 FORMAT('0','Aquifer top elevation in ft:')
    CALL DATPR(NC,NR,NL,CH,K,NZ)
    WRITE(4,446)
446 FORMAT('0','Aquifer base elevation in ft:')
    CALL DATPR(NC,NR,NL,BOT,K,NZ)
    WRITE(4,447)
447 FORMAT('0','Initial head elevation in ft:')
    CALL DATPR(NC,NR,NL,H,K,NZ)
    WRITE(4,448)
448 FORMAT('0','Transmissivity in I-Direction in gpd/ft:')
    NZ=2
    CALL DATPR(NC,NR,NL,T1,K,NZ)
    WRITE(4,449)
449 FORMAT('0','Transmissivity in J-Direction in gpd/ft:')
    CALL DATPR(NC,NR,NL,T2,K,NZ)
    IF(M11.EQ.2.AND.K.EQ.1)THEN
    WRITE(4,450)
450 FORMAT('0','Aquifer Artesian Storativity as a decimal:')
    NZ=3
    CALL DATPR(NC,NR,NL,S1,K,NZ)
    ENDIF
    IF(MAU.EQ.3)GOTO 454
    IF(MAU.EQ.1.AND.K.LT.3)GOTO 454
    IF(MAU.EQ.2.AND.K.GT.2)GOTO 454
    WRITE(4,451)
451 FORMAT('0','Aquitard vertical hydraulic conductivity'/1X,
   *'in gpd/sq ft:')
    NZ=3
    CALL DATPR(NC,NR,NL,P3,K,NZ)
    WRITE(4,452)
452 FORMAT('0','Aquitard thickness in ft:')
    NZ=1
    CALL DATPR(NC,NR,NL,THICKC,K,NZ)
454 IF(MAS.EQ.2)GOTO 456
    WRITE(4,455)
455 FORMAT('0','Aquitard storativity as a decimal:')
    NZ=3
    CALL DATPR(NC,NR,NL,S3,K,NZ)
456 WRITE(4,457)
457 FORMAT('0','Simulation factor in gpd/ft:')
    NZ=2
    CALL DATPR(NC,NR,NL,R,K,NZ)
    WRITE(4,458)
458 FORMAT('0','Aquifer specific yield factor as a decimal:')
    NZ=3
```

```
      CALL DATPR(NC,NR,NL,S2,K,NZ)
  459 CONTINUE
  460 DO 462 I=1,NP
      WRITE(4,461)I,IP(I),JP(I),KP(I)
  461 FORMAT('0','Production well number',I3,/1X,
     *'I-coordinate=',I3,'  J-coordinate=',I3,/1X,
     *'Unit or layer number=',I2)
  462 CONTINUE
      DO 464 I=1,NOBS
      WRITE(4,463)I,IO(I),JO(I),KO(I)
  463 FORMAT('0','Observation well number',I3,/1X,
     *'I-coordinate=',I3,'  J-coordinate=',I3,/1X,
     *'unit or layer number=',I2)
  464 CONTINUE
      DO 466 I=1,NRW
      WRITE(4,465)I,KR(I),IR(I),JR(I)
  465 FORMAT('0','Injection well number',I3,' in aquifer'/1X,
     *'unit or layer number',I3,' I-coordinate=',I3,/1X,
     *'J-coordinate=',I3)
  466 CONTINUE
  467 DO 705 ISTEP=1,NSTEPS
C
C     ENTER SINK/SOURCE STRENGTHS
C
  470 IF(ISTEP.GT.1)THEN
      WRITE(*,471)ISTEP
  471 FORMAT('0','Enter 1 if no sink/source geometry or'/1X,
     *'strength changes are required during time increment',I3,/1X,
     *'or 2 to change existing sink/source values'/1X,
     *'during this time increment. Enter 2 if aquitard'/1X,
     *'storativity is simulated.  '\)
      READ(*,*)NSS
      IF(NSS.NE.1.AND.NSS.NE.2)GOTO 470
      IF(NSS.EQ.1)GOTO 496
      ENDIF
  473 DO 479 N=1,NP
      I=IP(N)
      J=JP(N)
      K=KP(N)
  474 WRITE(*,475)N
  475 FORMAT('0','Production well number=',I2)
      WRITE(*,476)
  476 FORMAT(1X,'Enter real number production well discharge rate'/1X,
     *'during this time increment in gpm  '\)
      READ(*,*)QP(I,J,K)
      WRITE(4,477)N
  477 FORMAT('0','Production well number=',I3)
      WRITE(4,478)QP(I,J,K)
  478 FORMAT(1X,'Production well discharge rate in gpm=',F9.2)
      QP(I,J,K)=QP(I,J,K)*1440
  479 CONTINUE
      DO 485 N=1,NRW
```

```
      WRITE(*,481)N
  481 FORMAT(1X,'Injection well number',I3)
      I=IR(N)
      J=JR(N)
      K=KR(N)
      WRITE(*,482)ISTEP
  482 FORMAT(1X,'Enter real number injection well recharge'/1X,
     *'rate during time increment',I3,'  in gpm  '\)
      READ(*,*)QI(I,J,K)
      WRITE(4,484)N,QI(I,J,K)
  484 FORMAT(1X,'Recharge rate in injection well',I3,/1X,
     *'in gpm=',F9.2)
      QI(I,J,K)=QI(I,J,K)*1440
  485 CONTINUE
  486 WRITE(*,487)
  487 FORMAT('0','Enter 1 for uniform areal recharge'/1X,
     *'or 2 for no recharge  '\)
      READ(*,*)NUR
      IF(NUR.NE.1.AND.NUR.NE.2)GOTO 486
      IF(NUR.EQ.2)GOTO 492
      WRITE(*,488)
  488 FORMAT('0','Enter real number uniform areal recharge'/1X,
     *'rate in inches/year, usually 1E-2 to 48 inches/year  '\)
      READ(*,*)URR
      DO 490 J=1,NR
      DO 489 I=1,NC
      QR(I,J,1)=URR*DELX(I)*DELY(J)*1.71E-3
  489 CONTINUE
  490 CONTINUE
      WRITE(4,491)URR
  491 FORMAT(1X,'Areal recharge rate in inches per year=',F9.3)
  492 WRITE(*,493)
  493 FORMAT('0','Enter 1 to simulate mine, river, drain,'/1X,
     *'or evapotranspiration or 2 to continue program  '\)
      READ(*,*)LC2
      IF(LC2.NE.1.AND.LC2.NE.2)GOTO 492
      DO 495 J=1,NR
      DO 494 I=1,NC
      MQE(I,J,1)=0
  494 CONTINUE
  495 CONTINUE
      IF(LC2.EQ.2)GOTO 496
      CALL MINE(TIME,ISTEP,DELTA,DELX,DELY,THICK,SF1,SF2,
     *S1,S2,H,HO,T1,T2,P1,P2,QMD,BOT,MDM1,NDMN,MDM2,
     *MDM3,MDM4,LC1)
      CALL RIVER(R,RH,RD,MSB,MS1,MSN)
      CALL DRAIN(R,RH,RD,MDR,MD1,MDN)
      CALL EVAP(R,QP,QE,RD,ME1,RH,MQE,DELX,DELY,NC,NR)
C
C     SIMULATION OF AQUITARD STORATIVITY
C
  496 IF(MAS.EQ.1)THEN
```

```
      WRITE(*,497)
  497 FORMAT('0','Enter real number elapsed time after last'/1X,
     *'production well discharge rate change in days. A'/1X,
     *'complicated aquitard storativity function calculation'/1X,
     *'follows so be patient.  '\)
      READ(*,*)T1X
      ENDIF
      IF(MAS.NE.1)GOTO 502
      DO 501 K=1,NL
      IF(MUS.EQ.2.AND.K.EQ.1)GOTO 501
      DO 500 J=1,NR
      DO 499 I=1,NC
      T1X1=S3(I,J,K)*THICKC(I,J,K)/P3(I,J,K)
      T1X2=TIME(ISTEP)
      IF(T1X2.GT.T1X1)THEN
      RS(I,J,K)=0.0
      GOTO 499
      ENDIF
      F1=P3(I,J,K)*DELX(I)*DELY(J)
      F2=3.1416*P3(I,J,K)*T1X*THICKC(I,J,K)
      F3=3*THICKC(I,J,K)**2*S3(I,J,K)*7.48
      F4=(F2/F3)**.5
      F5=F1/(F4*THICKC(I,J,K))
      F6=P3(I,J,K)*T1X/(3*THICKC(I,J,K)**2*S3(I,J,K)*7.48)*
     *THICKC(I,J,K)
      F7=0.0
      DO 498 N=1,200
      F8=EXP(-N**2/F6)
      F7=F7+F8
  498 CONTINUE
      RS(I,J,K)=F5*(1+2*F7)
  499 CONTINUE
  500 CONTINUE
  501 CONTINUE
  502 IF(NSS.EQ.1)GOTO 511
  503 WRITE(*,504)
  504 FORMAT('0','Enter 1 to revise data base which varies'/1X,
     *'with time or 2 to continue program  '\)
      READ(*,*)RTDB
      IF(RTDB.NE.1.AND.RTDB.NE.2)GOTO 503
      IF(RTDB.EQ.1)GOTO 473
  505 IF(MF1.NE.1)GOTO 511
  506 WRITE(*,508)
  508 FORMAT('0','Enter 1 to cap flowing wells or'/1X,
     *'2 to continue flow  '\)
      READ(*,*)MF2
      IF(MF2.NE.1.AND.MF2.NE.2)GOTO 506
      IF(MF2.EQ.1)THEN
      DO 510 I=1,MNF
      L=MIF(I)
      N=MJF(I)
      K=MKF(I)
```

```
      R(L,N,K)=0.0
  510 CONTINUE
      ENDIF
C
C     SOLVE SIMULTANEOUS EQUATIONS AT NODES
C
  511 CALL SIMEQ(H,HO,SF1,Q,T1,T2,NC,NR,NL,DELTA,ISTEP,R,DL,RS,
     *MAS,DELX,DELY,MDY,SF2,MAP1,IMP,JMP,A7,A8,QMA,KUM,KLM,
     *ERROR,S1,S2,ALPHA,TIME,QDY,HD,E,ITER,WBE,NIC,QP,QI,QR,BOT,
     *CH,MFW,CELE,MSB,RD,RH,MDR,P1,P2,MF1,MDM1,MS1,MD1,ME1,
     *QF,QRD,QST,QDRAIN,QE,MUS,HSB,G,B,MQE,MF2,MAU)
  512 DO 513 IW=1,NOBS
      I=IO(IW)
      J=JO(IW)
      K=KO(IW)
      HTD(IW,ISTEP)=H(I,J,K)
  513 CONTINUE
      DO 515 J=1,NR
      DO 514 I=1,NC
      HD(I,J,ISTEP)=H(I,J,1)-HO(I,J,1)
  514 CONTINUE
  515 CONTINUE
C
C     DISPLAY RESULTS OF SOURCE/SINK COMPUTATIONS
C
      CALL RESLT(DELX,DELY,T1,T2,H,Q,TIME,ISTEP,
     *R,IP,JP,KP,RS,MAS,HO,SF1,DELTA,SF2,MDY,QDY,
     *MAP1,A7,A8,QMA,KUM,KLM,IMP,JMP,QMD,QRD,QST,QDRAIN,QF,
     *MF1,MDM1,MS1,MD1,MUS,HSB,CH,NL,MDM2)
C
C     PRINT AND EXPORT NODAL COMPUTATION RESULTS
C
  517 WRITE(*,518)
  518 FORMAT('0','Enter 1 for screen display of computation'/1X,
     *'nodal results or 2 for no display  '\)
      READ(*,*)NSCR
      IF(NSCR.NE.1.AND.NSCR.NE.2)GOTO 517
      IF(NSCR.EQ.2)GOTO 600
      DO 595 K=1,NL
      WRITE(*,520)
  520 FORMAT('0','Use Ctrl-Num Lock keys to'/1X,
     *'control screen scrolling')
      PAUSE
      WRITE(*,522)
  522 FORMAT('0','COMPUTATION NODAL RESULTS:')
      WRITE(*,530)TIME(ISTEP)
  530 FORMAT('0','TIME (DAYS)=',F12.5)
      WRITE(*,540)K
  540 FORMAT('0','HEAD OR DRAWDOWN (FT) IN UNIT OR LAYER',I2,':')
      WRITE(*,550)
  550 FORMAT('0','ROW                                COLUMN'/)
      N1=1
```

```
      N2=8
      IF(NC.LT.N2)N2=NC
  560 WRITE(*,570)(I,I=N1,N2)
  570 FORMAT(3X,8I9)
      DO  590 J=1,NR
      WRITE(*,580)J,(H(I,J,K),I=N1,N2)
  580 FORMAT(I3,8F9.2)
  590 CONTINUE
      IF(N2.EQ.NC)GOTO 595
      N1=N1+8
      N2=N2+8
      IF(NC.LT.N2)N2=NC
      GOTO 560
      PAUSE
  595 CONTINUE
  600 WRITE(*,610)
  610 FORMAT('0','Enter 1 for hard copy of computation'/1X,
     *'nodal results or 2 for no copy  '\)
      READ(*,*)NHR
      IF(NHR.NE.1.AND.NHR.NE.2)GOTO 600
      IF(NHR.EQ.2)GO TO 646
      DO 645 K=1,NL
      WRITE(4,612)
  612 FORMAT('0','COMPUTATION NODAL RESULTS:')
      WRITE(4,620)TIME(ISTEP)
  620 FORMAT('0','TIME(DAYS)=',F12.5)
      WRITE(4,630)K
  630 FORMAT('0','HEAD OR DRAWDOWN (FT) IN UNIT OR LAYER',I2,':')
      WRITE(4,640)
  640 FORMAT('0','ROW                              COLUMN'/)
      NZ=1
      CALL DATPR(NC,NR,NL,H,K,NZ)
  645 CONTINUE
  646 WRITE(*,647)
  647 FORMAT('0','Enter 1 to simulate partially penetrating'/1X,
     *'wells or well storage capacity, create head files,'/1X,
     *'create graphic format files, display head profiles,'/1X,
     *'or print and display time-head data. Some data'/1X,
     *'will be displayed on the screen only. Enter 2 to'/1X,
     *'continue program.  '\)
      READ(*,*)MGB1
      IF(MGB1.NE.1.AND.MGB1.NE.2)GOTO 646
      IF(MGB1.EQ.2)GOTO 705
C
C     SIMULATE PARTIAL PENETRATION AND WELL STORAGE
C     CAPACITY
C
  649 IF(NP.NE.0)THEN
      CALL PARPEN
      CALL WELSTOR(ISTEP,QAT)
      ENDIF
  650 WRITE(*,651)
  651 FORMAT('0','Enter 1 to create head files or 2'/1X,
```

```
     *'to continue program  '\)
      READ(*,*)MG1
      IF(MG1.NE.1.AND.MG1.NE.2)GOTO 650
      IF(MG1.EQ.2)GOTO 704
      DO 703 K=1,NL
  659 WRITE(*,660)K
  660 FORMAT('0','Enter 1 to create an ASCII file of comput-'/1X,
     *'ation nodal results by rows with one head on a line'/1X,
     *'for unit or layer',I2,' or 2 for no file  '\)
      READ(*,*)NFL
      IF(NFL.NE.1.AND.NFL.NE.2)GOTO 659
      IF(NFL.EQ.2)GOTO 704
      WRITE(*,670)
  670 FORMAT('0','Enter a valid computation nodal result'/1X,
     *'filename with extension DAT such as B:FILE2.DAT'/1X,
     *'in apostrophes. If apostrophes are ommitted,'/1X,
     *'a fatal error will occur.  '\)
      READ(*,*)FNAME
      OPEN(6,FILE=FNAME,STATUS='NEW')
      DO 690 J=1,NR
      DO 685 I=1,NC
      WRITE(6,680)H(I,J,K)
  680 FORMAT(1X,F9.2)
  685 CONTINUE
  690 CONTINUE
      CLOSE (6)
  703 CONTINUE
C
C     CREATE FILE FOR GRAPHIC AND HEAD PROFILE DISPLAYS
C
  704 CALL XYZDT(H,NC,NR,DELX,DELY)
      CALL PROFIL(H,NL)
  705 CONTINUE
C
C     PRINT AND EXPORT TIME-HEAD COMPUTATION RESULTS
C
      IF(NOBS.NE.0.OR.MGB1.EQ.1)THEN
  710 WRITE(*,720)
  720 FORMAT('0','Enter 1 for screen display of computation'/1X,
     *'time-head results or 2 for no display  '\)
      READ(*,*)NTD
      IF(NTD.NE.1.AND.NTD.NE.2)GOTO 710
      IF(NTD.EQ.2)GOTO 790
      WRITE(*,730)
  730 FORMAT('0','Use Ctrl-Num Lock keys to control screen'/1X,
     *'scrolling')
      DO 770 IW=1,NOBS
      WRITE(*,740)IW
  740 FORMAT('0','Observation Well Number',I3)
      WRITE(*,750)
  750 FORMAT('0',' Time (Days)      Head (Ft)')
      DO 765 ISTEP=1,NSTEPS
      WRITE(*,760)TIME(ISTEP),HTD(IW,ISTEP)
```

```
760 FORMAT(1X,F12.5,4X,F10.2)
765 CONTINUE
    PAUSE
770 CONTINUE
790 WRITE(*,800)
800 FORMAT('0','Enter 1 for hard copy of computation'/1X,
   *'time-head results or 2 for no copy  '\)
    READ(*,*)NTDH
    IF(NTDH.NE.1.AND.NTDH.NE.2)GOTO 790
    IF(NTDH.EQ.2)GOTO 860
    DO 840 IW=1,NOBS
    WRITE(4,810)IW
810 FORMAT('0','Observation Well Number',I3)
    WRITE(4,820)
820 FORMAT('0',' Time (Days)      Head (Ft)')
822 DO 835 ISTEP=1,NSTEPS
    WRITE(4,830)TIME(ISTEP),HTD(IW,ISTEP)
830 FORMAT(1X,F12.5,4X,F10.2)
835 CONTINUE
840 CONTINUE
    ENDIF
860 CLOSE (4)
    IF(NOBS.NE.0)THEN
870 WRITE(*,880)
880 FORMAT('0','Enter 1 to create an ASCII file of time-'/1X,
   *'head values by rows with time and head on a line'/1X,
   *'separated by a comma or 2 for no file  '\)
    READ(*,*)NFT
    IF(NFT.NE.1.AND.NFT.NE.2)GOTO 870
    IF(NFT.EQ.2)GOTO 950
    WRITE(*,890)
890 FORMAT('0','Enter integer number of observation'/1X,
   *'wells for which files are to be created  '\)
    READ(*,*)NFO
    DO 930 I=1,NFO
    WRITE(*,900)
900 FORMAT('0','Enter a valid computed time-head'/1X,
   *'filename with extension DAT such as B:FILE3.DAT'/1X,
   *'in apostrophes. If apostrophes are ommitted, a fatal'/1X,
   *'error will occur.  '\)
    READ(*,*)FNAME
    OPEN(6,FILE=FNAME,STATUS='NEW')
    WRITE(*,910)
910 FORMAT(1X,'Enter integer number of observation well  '\)
    READ(*,*)IW
    DO 925 ISTEP=1,NSTEPS
    WRITE(6,920)TIME(ISTEP),HTD(IW,ISTEP)
920 FORMAT(1X,F12.5,',',F10.2)
925 CONTINUE
930 CONTINUE
    ENDIF
    CLOSE (6)
950 END
```

```
C
      SUBROUTINE DATPR(NC,NR,NL,HHE,K,NZ)
C
C     THIS SUBROUTINE PRINTS A HARD COPY OF
C     VARIABLE VALUES AS REQUESTED
C
C     METACOMMANDS
C
$LARGE
C
C     DECLARATIONS
C
      DIMENSION HHE(20,20,4)
C
      OPEN(5,FILE='LPT1:')
      N1=1
      N2=8
      IF(NC.LT.N2)N2=NC
  110 WRITE(5,120)(I,I=N1,N2)
  120 FORMAT(3X,8I9)
      DO 140 J=1,NR
      IF(NZ.EQ.2)THEN
      WRITE(5,122)J,(HHE(I,J,K),I=N1,N2)
  122 FORMAT(I3,8F9.0)
      GOTO 140
      ENDIF
      IF(NZ.EQ.3)THEN
      WRITE(5,124)J,(HHE(I,J,K),I=N1,N2)
  124 FORMAT(I3,8F9.6)
      GOTO 140
      ENDIF
      IF(NZ.EQ.1)THEN
      WRITE(5,130)J,(HHE(I,J,K),I=N1,N2)
  130 FORMAT(I3,8F9.2)
      ENDIF
  140 CONTINUE
      IF(N2.EQ.NC)GOTO 150
      N1=N1+8
      N2=N2+8
      IF(NC.LT.N2)N2=NC
      GOTO 110
  150 CLOSE (5)
      END
```

```
C
      SUBROUTINE DRAIN(R,RH,RD,MDR,MD1,MDN)
C
C      THIS SUBROUTINE ENTERS AND DISPLAYS DRAIN SINK
C      GEOMETRY AND STRENGTH DATA BASE
C
C      DECLARATIONS
C
      DIMENSION R(20,20,4),RH(20,20,1),RD(20,20,1),MDR(20,20,1)
C
    5 WRITE(*,7)
    7 FORMAT('0','Enter 1 to simulate drains in aquifer'/1X,
     *'unit or layer 1 or 2 to continue program  '\)
      READ(*,*)MD1
      IF(MD1.NE.1.AND.MD1.NE.2)GOTO 5
      IF(MD1.EQ.1)THEN
      OPEN(5,FILE='LPT1:')
   10 WRITE(*,20)
   20 FORMAT('0','Enter integer number of drain nodes in'/1X,
     *'uppermost aquifer unit or layer  '\)
      READ(*,*)MDN
      DO 200 N=1,MDN
      WRITE(*,30)N
   30 FORMAT(1X,'Drain node number=',I3)
      WRITE(5,40)N
   40 FORMAT(1X,'Drain node number=',I3)
      WRITE(*,50)
   50 FORMAT(1X,'Enter integer I-coordinate of drain node  '\)
      READ(*,*)I
      WRITE(5,60)I
   60 FORMAT(1X,'I-coordinate of drain node=',I3)
      WRITE(*,70)
   70 FORMAT(1X,'Enter integer J-coordinate of drain node  '\)
      READ(*,*)J
      WRITE(5,80)J
   80 FORMAT(1X,'J-coordinate of drain node=',I3)
      WRITE(*,90)
   90 FORMAT(1X,'Enter real number drain top elevation'/1X,
     *'in ft  '\)
      READ(*,*)RH(I,J,1)
      RD(I,J,1)=RH(I,J,1)
      WRITE(5,100)RH(I,J,1)
  100 FORMAT(1X,'Drain top elevation in ft=',F9.2)
      WRITE(*,110)
  110 FORMAT(1X,'Enter real number thickness of backfill'/1X,
     *'around drain in ft  '\)
```

```
      READ(*,*)DTHICK
      WRITE(5,120)DTHICK
120 FORMAT(1X,'Thickness of backfill around drain'/1X,
   *'in ft=',F6.2)
      WRITE(*,130)
130 FORMAT('0','The effective hydraulic conductivity'/1X,
   *'of backfill around drain is affected by size and'/1X,
   *'frequency of openings in drain tile, chemical'/1X,
   *'precipitation around tile, and hydraulic conductivities'/1X,
   *'of backfill or low permeability bed in open drain'/)
      WRITE(*,140)
140 FORMAT(1X,'Enter real number drain backfill effective'/1X,
   *'hydraulic conductivity in gpd/sq ft, usually ranges from'/1X,
   *'1E-2 to 1000 gpd/sq ft  '\)
      READ(*,*)DHC
      WRITE(5,150)DHC
150 FORMAT(1X,'Effective hydraulic conductivity of'/1X,
   *'backfill around drain in gpd/sq ft=',F8.2)
      WRITE(*,160)
160 FORMAT(1X,'Enter real number drain diameter in inches  '\)
      READ(*,*)DDI
      WRITE(5,170)DDI
170 FORMAT(1X,'Drain diameter in inches=',F7.2)
      DDI=DDI/12
      WRITE(*,180)
180 FORMAT(1X,'Enter real number length of drain'/1X,
   *'simulated at node in ft  '\)
      READ(*,*)DLE
      WRITE(5,190)DLE
190 FORMAT(1X,'Length of drain simulated at node in ft=',F9.2)
      R(I,J,1)=(DHC/DTHICK)*(2*DTHICK+DDI)*DLE
      MDR(I,J,1)=1
200 CONTINUE
210 WRITE(*,220)
220 FORMAT('0','Enter 1 to revise drain data base'/1X,
   *'or 2 to continue program  '\)
      READ(*,*)MD2
      IF(MD2.NE.1.AND.MD2.NE.2)GOTO 210
      IF(MD2.EQ.1)GOTO 10
      CLOSE (5)
      ENDIF
230 END
```

```
C
      SUBROUTINE EVAP(R,QP,QE,RD,ME1,RH,MQE,DELX,DELY,NC,NR)
C
C     THIS SUBROUTINE ENTERS AND DISPLAYS AREAL
C     EVAPOTRANSPIRATION SINK GEOMETRY AND STRENGTH DATA
C     BASE
C
C     DECLARATIONS
C
      DIMENSION R(20,20,4),QP(20,20,4),QE(20,20,1),RD(20,20,1),
     *RH(20,20,1),MQE(20,20,1),DELX(21),DELY(21)
C
    2 WRITE(*,3)
    3 FORMAT('0','Enter 1 to simulate evapotranspiration'/1X,
     *'in uppermost aquifer unit or layer under water'/1X,
     *'table conditions or 2 to continue program  '\)
      READ(*,*)ME1
      IF(ME1.NE.1.AND.ME1.NE.2)GOTO 2
      IF(ME1.EQ.1)THEN
      OPEN(5,FILE='LPT1:')
    5 WRITE(*,10)
   10 FORMAT('0','Enter real number potential(pan)'/1X,
     *'evapotranspiration rate in inches/year,usually'/1X,
     *'ranges from 12 to 120 inches/year  '\)
      READ(*,*)ET
      WRITE(5,20)ET
   20 FORMAT(1X,'Potential evapotranspiration rate'/1X,
     *'in inches/year=',F7.2)
      ET=ET*1.71E-3
      WRITE(*,22)
   22 FORMAT(1X,'Enter real number elevation below land'/1X,
     *'surface at which evapotranspiration ceases in ft  '\)
      READ(*,*)EVSTOP
      WRITE(5,24)EVSTOP
   24 FORMAT(1X,'Elevation below land surface at which'/1X,
     *'evapotranspiration ceases in ft=',F9.2)
      WRITE(*,26)
   26 FORMAT(1X,'Enter real number land surface elevation'/1X,
     *'in ft  '\)
      READ(*,*)SLS
      WRITE(5,28)SLS
   28 FORMAT(1X,'Land surface elevation in ft=',F9.2)
      DO 40 J=1,NR
      DO 30 I=1,NC
      IF(R(I,J,1).NE.0.0)GOTO 30
      RD(I,J,1)=EVSTOP
```

```
      RH(I,J,1)=SLS
      QE(I,J,1)=ET*DELX(I)*DELY(J)
      R(I,J,1)=QE(I,J,1)/(RH(I,J,1)-RD(I,J,1))
      MQE(I,J,1)=1
30    CONTINUE
40    CONTINUE
50    WRITE(*,60)
60    FORMAT('0','Enter 1 to revise evapotranspiration'/1X,
     *'data base or 2 to continue program   '\)
      READ(*,*)MV1
      IF(MV1.NE.1.AND.MV1.NE.2)GOTO 50
      IF(MV1.EQ.1)GOTO 5
      CLOSE (5)
      ENDIF
70    END
```

```
C
      SUBROUTINE FILEI(NC,NR,NSTEPS,NP,NOBS,ERROR,WBE,TIME,
     *DELTA,IP,JP,KP,IO,JO,KO,NL,H,THICK,P1,P2,T1,T2,P3,
     *THICKC,SF1,S1,R,DELX,DELY,Q,P4,S3,MAS,S2,RS,DL,MF1,MNF,
     *MIF,MJF,MKF,MFW,CELE,SF2,CH,BOT,NRW,IR,JR,KR,MUS,HSB,
     *MAL,MAU,KFI,MDY,LC1,QI,QR,QMD,MSB,MDR,QE,QP,
     *QF,QST,QDRAIN,RH,RD)
C
C     THIS SUBROUTINE IMPORTS DATA BASE
C
C     METACOMMANDS
C
$LARGE
C
C     DECLARATIONS
C
      CHARACTER*14FNAME
      DIMENSION TIME(50),DELTA(50),IP(50),JP(50),KP(4),IO(25),
     *JO(25),KO(4),H(20,20,4),THICK(20,20,4),
     *P1(20,20,4),P2(20,20,4),T1(20,20,4),T2(20,20,4),P3(20,20,4),
     *THICKC(20,20,4),SF1(20,20,4),S1(20,20,4),R(20,20,4),
     *DELX(21),DELY(21),Q(20,20,4),DL(20,20,4),P4(20,20,4),
     *S3(20,20,4),RS(20,20,4),S2(20,20,4),MIF(50),MJF(50),MKF(50),
     *MFW(20,20,4),CELE(20,20,4),SF2(20,20,4),CH(20,20,4),
     *BOT(20,20,4),QI(20,20,4),QR(20,20,1),QMD(20,20,4),
     *QF(20,20,4),QRD(20,20,4),IR(50),JR(50),KR(50),
     *QE(20,20,1),QP(20,20,4),MSB(20,20,1),
     *MDR(20,20,1),QST(20,20,1),QDRAIN(20,20,1),RH(20,20,1),
     *RD(20,20,1)
   10 WRITE(*,20)
   20 FORMAT('0','Enter a valid data base specification'/1X,
     *'filename with extension DAT of ASCII file to be'/1X,
     *'imported such as B:file4.dat in apostrophes. If the'/1X,
     *'apostrophes are ommitted a fatal error will occur.  '\)
      READ(*,*)FNAME
      OPEN(6,FILE=FNAME)
      READ(6,30)NC,NR,NSTEPS,NP,NOBS,NL,MAS,MF1,MNF,NRW,MUS
   30 FORMAT(1X,11I3)
      READ(6,32)MAL,MAU,KFI,MDY,LC1
   32 FORMAT(1X,5I3)
      READ(6,40)ERROR,WBE
   40 FORMAT(1X,F9.2,F6.2)
      DO 60 I=1,NSTEPS
      READ(6,50)TIME(I),DELTA(I)
   50 FORMAT(1X,2F14.5)
   60 CONTINUE
```

```
      DO 80 I=1,NP
      READ(6,70)IP(I),JP(I),KP(I)
   70 FORMAT(1X,3I3)
   80 CONTINUE
      DO 84 I=1,MNF
      READ(6,82)MIF(I),MJF(I),MKF(I)
   82 FORMAT(1X,3I3)
   84 CONTINUE
      DO 88 I=1,NRW
      READ(6,86)IR(I),JR(I),KR(I)
   86 FORMAT(1X,3I3)
   88 CONTINUE
      DO 100 I=1,NOBS
      READ(6,90)IO(I),JO(I),KO(I)
   90 FORMAT(1X,3I3)
  100 CONTINUE
      READ(6,92)HSB
   92 FORMAT(1X,F9.2)
      DO 150 K=1,NL
      DO 140 J=1,NR
      DO 130 I=1,NC
      READ(6,110)H(I,J,K),THICK(I,J,K),P1(I,J,K),P2(I,J,K),
     *P3(I,J,K),THICKC(I,J,K),P4(I,J,K),MFW(I,J,K)
  110 FORMAT(1X,7F9.2,I3)
      READ(6,120)T1(I,J,K),T2(I,J,K),SF1(I,J,K),S1(I,J,K),R(I,J,K),
     *S3(I,J,K),RS(I,J,K),S2(I,J,K),CELE(I,J,K),SF2(I,J,K),
     *CH(I,J,K),BOT(I,J,K)
  120 FORMAT(1X,3F12.2,F12.8,F12.2,F12.8,F12.2,F12.8,F10.2,
     *F12.2,2F10.2)
      READ(6,125)Q(I,J,K),DL(I,J,K),QI(I,J,K),QMD(I,J,K),
     *QP(I,J,K),QF(I,J,K)
  125 FORMAT(1X,6F12.2)
  130 CONTINUE
  140 CONTINUE
  150 CONTINUE
      DO 158 J=1,NR
      DO 156 I=1,NC
      READ(6,152)QR(I,J,1),QE(I,J,1),QST(I,J,1),QDRAIN(I,J,1),
     *RH(I,J,1),RD(I,J,1),MSB(I,J,1),MDR(I,J,1)
  152 FORMAT(1X,4F12.2,2F9.2,2I3)
  156 CONTINUE
  158 CONTINUE
      DO 170 I=1,NC+1
      READ(6,160)DELX(I)
  160 FORMAT(1X,F9.2)
  170 CONTINUE
      DO 190 J=1,NR+1
      READ(6,180)DELY(J)
  180 FORMAT(1X,F9.2)
  190 CONTINUE
      CLOSE (6)
      END
```

```
C
      SUBROUTINE FILEO(NC,NR,NSTEPS,NP,NOBS,ERROR,WBE,TIME,
     *DELTA,IP,JP,KP,IO,JO,KO,NL,H,THICK,P1,P2,T1,T2,P3,
     *THICKC,SF1,S1,R,DELX,DELY,Q,P4,S3,MAS,S2,RS,DL,MF1,
     *MNF,MIF,MJF,MKF,MFW,CELE,SF2,CH,BOT,NRW,IR,JR,KR,MUS,HSB,
     *MAL,MAU,KFI,MDY,LC1,QI,QR,QMD,MSB,MDR,QE,QP,
     *QF,QST,QDRAIN,RH,RD)
C
C     THIS SUBROUTINE EXPORTS DATA BASE
C
C     METACOMMANDS
C
$LARGE
C
C     DECLARATIONS
C
      CHARACTER*14FNAME
      DIMENSION TIME(50),DELTA(50),IP(50),JP(50),KP(4),
     *IO(25),JO(25),KO(4),H(20,20,4),THICK(20,20,4),
     *P1(20,20,4),P2(20,20,4),T1(20,20,4),T2(20,20,4),
     *P3(20,20,4),THICKC(20,20,4),SF1(20,20,4),S1(20,20,4),
     *R(20,20,4),DELX(21),DELY(21),Q(20,20,4),DL(20,20,4),
     *P4(20,20,4),S3(20,20,4),RS(20,20,4),S2(20,20,4),MIF(50),
     *MJF(50),MKF(50),MFW(20,20,4),CELE(20,20,4),SF2(20,20,4),
     *CH(20,20,4),BOT(20,20,4),QI(20,20,4),QR(20,20,4),
     *QMD(20,20,4),QF(20,20,4),QRD(20,20,4),IR(50),JR(50),KR(50),
     *QE(20,20,1),QP(20,20,4),MSB(20,20,1),
     *MDR(20,20,1),QST(20,20,1),QDRAIN(20,20,1),RH(20,20,1),
     *RD(20,20,1)
      WRITE(*,324)
  324 FORMAT('0','Enter a valid data base specification'/1X,
     *'filename with extension DAT of ASCII file to be'/1X,
     *'exported such as B:FILE1.DAT in apostrophes. If the'/1X,
     *'apostrophes are ommitted a fatal error will occur.  '\)
      READ(*,*)FNAME
      OPEN(6,FILE=FNAME,STATUS='NEW')
      WRITE(6,328)NC,NR,NSTEPS,NP,NOBS,NL,MAS,MF1,MNF,NRW,MUS
  328 FORMAT(1X,11I3)
      WRITE(6,329)MAL,MAU,KFI,MDY,LC1
  329 FORMAT(1X,5I3)
      WRITE(6,330)ERROR,WBE
  330 FORMAT(1X,F9.2,F6.2)
      DO 332 I=1,NSTEPS
      WRITE(6,331)TIME(I),DELTA(I)
  331 FORMAT(1X,2F14.5)
  332 CONTINUE
```

```
      DO 336 I=1,NP
      WRITE(6,334)IP(I),JP(I),KP(I)
  334 FORMAT(1X,3I3)
  336 CONTINUE
      DO 342 I=1,MNF
      WRITE(6,340)MIF(I),MJF(I),MKF(I)
  340 FORMAT(1X,3I3)
  342 CONTINUE
      DO 344 I=1,NRW
      WRITE(6,343)IR(I),JR(I),KR(I)
  343 FORMAT(1X,3I3)
  344 CONTINUE
      DO 346 I=1,NOBS
      WRITE(6,345)IO(I),JO(I),KO(I)
  345 FORMAT(1X,3I3)
  346 CONTINUE
      WRITE(6,348)HSB
  348 FORMAT(1X,F9.2)
      DO 356 K=1,NL
      DO 354 J=1,NR
      DO 353 I=1,NC
      WRITE(6,350)H(I,J,K),THICK(I,J,K),P1(I,J,K),
     *P2(I,J,K),P3(I,J,K),THICKC(I,J,K),P4(I,J,K),MFW(I,J,K)
  350 FORMAT(1X,7F9.2,I3)
      WRITE(6,351)T1(I,J,K),T2(I,J,K),SF1(I,J,K),S1(I,J,K),
     *R(I,J,K),S3(I,J,K),RS(I,J,K),S2(I,J,K),CELE(I,J,K),
     *SF2(I,J,K),CH(I,J,K),BOT(I,J,K)
  351 FORMAT(1X,3F12.2,F12.8,F12.2,F12.8,F12.2,F12.8,F10.2,
     *F12.2,2F10.2)
      WRITE(6,352)Q(I,J,K),DL(I,J,K),QI(I,J,K),QMD(I,J,K),
     *QP(I,J,K),QF(I,J,K)
  352 FORMAT(1X,6F12.2)
  353 CONTINUE
  354 CONTINUE
  356 CONTINUE
      DO 359 J=1,NR
      DO 358 I=1,NC
      WRITE(6,357)QR(I,J,1),QE(I,J,1),QST(I,J,1),
     *QDRAIN(I,J,1),RH(I,J,1),RD(I,J,1),MSB(I,J,1),
     *MDR(I,J,1)
  357 FORMAT(1X,4F12.2,2F9.2,2I3)
  358 CONTINUE
  359 CONTINUE
      DO 361 I=1,NC+1
      WRITE(6,360)DELX(I)
  360 FORMAT(1X,F9.2)
  361 CONTINUE
      DO 364 J=1,NR+1
      WRITE(6,362)DELY(J)
  362 FORMAT(1X,F9.2)
  364 CONTINUE
      CLOSE (6)
      END
```

```
C
      SUBROUTINE FWEL(MNF,MFW,MIF,MJF,CELE,FWR,FWC,
     *T1,T2,DELX,DELY,R,MF1,MKF)
C
C     THIS SUBROUTINE ENTERS AND DISPLAYS FLOWING
C     WELL GEOMETRY
C
C     DECLARATIONS
C
      DIMENSION MFW(20,20,4),MIF(50),MJF(50),CELE(20,20,4),
     *T1(20,20,4),T2(20,20,4),DELX(21),DELY(21),R(20,20,4),
     *FWR(50),FWC(50),MKF(50)
C
   10 WRITE(*,20)
   20 FORMAT('0','Enter 1 to simulate flowing wells or 2'/1X,
     *'to continue data base specification. Springs'/1X,
     *'may be simulated as flowing wells. It is assumed that'/1X,
     *'water levels in the flowing well do not decline'/1X,
     *'below the top of the flowing well casing   '\)
      READ(*,*)MF1
      IF(MF1.NE.1.AND.MF1.NE.2)GOTO 10
      IF(MF1.EQ.1)THEN
      OPEN(5,FILE='LPT1:')
   25 WRITE(*,30)
   30 FORMAT('0','Enter integer number of flowing'/1X,
     *'wells (must be <51)   '\)
      READ(*,*)MNF
      DO 140 I=1,MNF
      WRITE(*,40)I
   40 FORMAT('0','Flowing well number=',I3)
      WRITE(*,42)
   42 FORMAT(1X,'Enter integer aquifer unit or layer'/1X,
     *'number in which flowing well is open   '\)
      READ(*,*)MKF(I)
      WRITE(*,50)
   50 FORMAT(1X,'Enter integer number I-coordinate'/1X,
     *'of flowing well   '\)
      READ(*,*)MIF(I)
      WRITE(*,60)
   60 FORMAT(1X,'Enter integer number J-coordinate'/1X,
     *'of flowing well   '\)
      READ(*,*)MJF(I)
      WRITE(*,70)
   70 FORMAT(1X,'Enter real number flowing well'/1X,
     *'effective radius in ft   '\)
      READ(*,*)FWR(I)
```

```
      WRITE(*,80)
   80 FORMAT(1X,'Enter real number flowing well'/1X,
     *'top of casing elevation in ft  '\)
      READ(*,*)FWC(I)
      WRITE(5,90)I,MKF(I),MIF(I),MJF(I)
   90 FORMAT(1X,'Flowing well',I3,' in aquifer unit or'/1X,
     *'layer number',I3,' I-coordinate=',I3,' J-coordinate=',I3)
      WRITE(5,100)FWR(I)
  100 FORMAT(1X,'Flowing well effective radius in ft=',F6.3)
      WRITE(5,110)FWC(I)
  110 FORMAT(1X,'Flowing well top of casing'/1X,
     *'elevation in ft=',F9.2)
      K=MKF(I)
      L=MIF(I)
      N=MJF(I)
      MFW(L,N,K)=1
      CELE(L,N,K)=FWC(I)
      TAV=(T1(L,N,K)*T2(L,N,K))**.5
      DAV=0.28*((T1(L,N,K)/T2(L,N,K))**.5*DELX(L)**2+
     *(T2(L,N,K)/T1(L,N,K))**.5*DELY(N)**2)**.5/
     *((T1(L,N,K)/T2(L,N,K))**.25+(T2(L,N,K)/T1(L,N,K))**.25)
      DF=DAV/FWR(I)
      R(L,N,K)=2*3.1416*TAV/ALOG(DF)
      WRITE(*,120)R(L,N,K)
  120 FORMAT(1X,'Flowing well simulation factor in gpd/ft=',F9.2)
      WRITE(5,130)R(L,N,K)
  130 FORMAT(1X,'Flowing well simulation factor in gpd/ft=',F9.2)
  140 CONTINUE
  150 WRITE(*,160)
  160 FORMAT('0','Enter 1 to revise flowing well data base or'/1X,
     *'2 to continue program  '\)
      READ(*,*)MF2
      IF(MF2.NE.1.AND.MF2.NE.2)GOTO 150
      IF(MF2.EQ.1)GOTO 25
      CLOSE (5)
      ENDIF
      END
```

```
C
      SUBROUTINE INFO1
C
C     THIS SUBROUTINE PROVIDES PROGRAM INFORMATION
C
      WRITE(*,10)
   10 FORMAT(1X,
     *'Program:          GWFL3D'/1X,
     *'Date:             July 1988'/1X,
     *'Author:           W.C.Walton modified the Prickett-Lonnquist'/1X,
     *'                  (1971) program'/1X,
     *'Version:          IBM/PC Microsoft FORTRAN77 v3.20 or 4.00'/1X,
     *'                  Copyright 1988 Lewis Publishers, Inc.'/1X
     *'Description:      Modular program for a one- two- or three-'/1X,
     *'                  dimensional finite-difference model'/1X,
     *'                  simulating groundwater flow with constant'/1X,
     *'                  density through saturated aquifer systems.'/1X,
     *'                  Hydraulic characteristics may vary horizontally'/1X,
     *'                  and vertically.'/1X,
     *'References:       Prickett,T.A. and C.G.Lonnquist.1971.Selected'/1X,
     *'                  digital computer techniques for groundwater'/1X,
     *'                  resources evaluation.Illinois State Water'/1X,
     *'                  Survey.Bull.55.62 p.')
      WRITE(*,12)
   12 FORMAT(1X,
     *'                  Trescott,P.C.,G.F.Pinder,and S.P.Larson.1976.'/1X,
     *'                  Finite-difference model for aquifer simulation'/1X,
     *'                  in two dimensions with results of numerical'/1X,
     *'                  experiments.Techniques of Water Resources'/1X,
     *'                  Investigations.U.S.Geol. Surv.Chapt.C1,Book 7'/1X,
     *'                  116 p.')
      PAUSE
      WRITE(*,14)
   14 FORMAT(1X,
     *'Trademarks:       IBM/PC International Business Machines Corp.'/1X,
     *'                  Microsoft FORTRAN77 Microsoft Corp.')
C     Definitions:
C
C     Name               Description
C
C     AA                 Used to calculate Peaceman-Rachford arrays
C     ALD                Leakage rate from above aquifer in gpd
C     ALU                Leakage rate from below aquifer in gpd
C     B                  Peaceman-Rachford B array
C     BB                 Used to calculate Peaceman-Rachford arrays
C     CC                 Used to calculate Peaceman-Rachford arrays
```

```
C     CELE        Flowing well top of casing elevation in ft
C     DD          Used to calculate Peaceman-Rachford arrays
C     DDI         Drain diameter in ft
C     DLE         Drain length in ft
C     DHC         Drain backfill effective hydraulic conductivity
C                 in gpd/sq ft
C     DPEN        Depth of penetration of mine below top
C                 of aquifer in ft
C     DTHICK      Thickness of backfill around drain in ft
C     DELTA       Time increment in days
C     DELX        Grid spacing in I-direction in ft
C     DELY        Grid spacing in J-direction in ft
C     DIFF        Water balance as a percentage difference
C                 between inflow and outflow
C     DLP         Distance from aquifer top to production well
C                 base in ft
C     DDP         Distance from aquifer top to production well
C                 screen top in ft
C     DLO         Distance from aquifer top to observation well
C                 base in ft
C     DDO         Distance from aquifer top to observation well
C                 screen top in ft
C     DEL         Time after pumping started in minutes
C     DDT         Drawdown at production well in ft
C     DDD         Drawdown at observation well in ft
C     DTOT        Total drawdown with partial penetration in ft
C     DPP         Drawdown due to partial penetration in ft
C     EMF         Mine floor elevation in ft
C     EVAP        Potential evapotranspiration rate in in/yr
C     EVSTOP      Elevation below land surface at which
C                 evapotranspiration ceases in ft
C     ERROR       Grid head tolerance error in ft
C     FWC         Flowing well effective radius in ft
C     FWR         Flowing well top of casing elevation in ft
C     G           Peaceman-Rachford G array
C     H           Head at end of time increment in ft
C     HCBM        Mine backfill material hydraulic conductivity
C                 in gpd/sq ft
C     HMA         Composite water level or drawdown in
C                 multi-aquifer production well in ft
C     HCE         Nodal head tolerance error in ft
C     HTD         Head at a particular observation well and
C                 time in ft
C     HO          Head at beginning of time increment in ft
C     HSB         Head in sourcebed above uppermost unit in ft
C     I           Index variable
C     ISTEP       Index variable
C     ITER        Iteration number
C     IO          I-coordinate of observation well
C     IVG1,IVG2   I-coordinates of grid block corners
C     JO          J-coordinate of observation well
C     IP          I-coordinate of production well
```

```
C     JP              J-coordinate of production well
C     IR              I-coordinate of injection well
C     JR              J-coordinate of injection well
C     IMP             I-coordinate of multi-aquifer production well
C     JMP             J-coordinate of multi-aquifer production well
C     J               Index variable
C     JVG1,JVG2       J-coordinates of grid block corners
C     INODE           I-coordinate of profile node
C     JNODE           J-coordinate of profile node
C     K               Index variable
C     KO              Number of aquifer unit or layer in which observation
C                     well is open
C     KP              Number of aquifer unit or layer in which production
C                     well is open
C     KLM             Lowermost aquifer unit or layer open to multi-
C                     aquifer production well
C     KUM             Uppermost aquifer unit or layer open to multi-
C                     aquifer production well
C     KR              Number of aquifer unit or layer in which
C                     injection well is open
C     L               Index variable
C     MNF             Number of flowing wells
C     MIF             I-coordinate of flowing well
C     MJF             J-coordinate of flowing well
C     MKF             Aquifer unit or layer in which flowing
C                     well is open
C     MDMN            Number of mine nodes
C     MSN             Number of streambed nodes
C     MSB             Identifies streambed node
C     MFW             Identifies flowing well node
C     MQE             Identifies evapotranspiration nodes
C     MDN             Number of drain nodes
C     MDR             Identifies drain node
C     MUS             Identifies leakage from sourcebed above
C                     uppermost unit
C     N               Index variable
C     NC              Number of grid columns
C     NL              Number of aquifer units or layers
C     NP              Number of production wells
C     NGB             Number of grid blocks
C     NOBS            Number of observation wells
C     NR              Number of grid rows
C     NSTEPS          Number of time increments
C     NODE            Number of nodes on profile
C     NRW             Number of injection wells
C     P1              Aquifer hydraulic conductivity in I-direction
C                     between I,J and I,J+1 in gpd/sq ft
C     P2              Aquifer hydraulic conductivity in J-direction
C                     between I,J and I+1,J in gpd/sq ft
C     P3              Aquitard vertical hydraulic conductivity
C                     in gpd/sq ft
C     P4              Aquifer vertical hydraulic conductivity
```

```
C                in gpd/sq ft
C     PUMP       Component of flow from discharge
C                units in gpd
C     Q          Production well discharge rate in gpd
C     QE         Groundwater evapotranspiration rate in gpd
C     QMD        Mine drift dewatering rate in gpd
C     QI         Injection well recharge rate in gpd
C     QF         Flowing well discharge rate in gpd
C     QR         Uniform areal recharge rate in gpd
C     QMA        Multi-aquifer production well discharge rate
C                in gpd
C     QDY        Gravity yield rate in gpd
C     R          Simulation factor in gpd/ft
C     RATIO      Ratio of aquifer vertical and horizontal
C                hydraulic conductivity
C     RH         Stream surface or drain base elevation or
C                land surface elevation in ft
C     RD         Streambed base elevation or depth below
C                which evapotranspiration ceases in ft
C     RW         Production well radius in ft
C     RWO        Distance between production and observation
C                wells in ft
C     RC         Discharge pipe-column radius in ft
C     RS         Simulation factor in gpd/ft
C     S1         Aquifer storativity as a decimal
C     S3         Aquitard artesian storativity as a decimal
C     S2         Aquifer specific yield as a decimal
C     SBW        Streambed width in ft
C     SBL        Streambed length in ft
C     SBT        Streambed thickness in ft
C     SBH        Streambed vertical hydraulic conductivity
C                in gpd/sq ft
C     STOR1      Mine backfill materials storativity as a decimal
C     STOR2      Mine backfill materials specific yield
C                as a decimal
C     SF2        Storage factor for water table conditions in gal/ft
C     STOR       Average aquifer storativity as a decimal
C     SF1        Storage factor for artesian conditions in gal/ft
C     STORE      Component of flow from aquifer storage
C                in gpd
C     STORE1     Component of flow from aquitard storage
C                in gpd
C     THICK      Aquifer thickness in ft
C     THICKC     Aquitard thickness in ft
C     TIME       Time after simulation started in days
C     T1         Aquifer transmissivity in I-direction between
C                I,J and I,J+1 in gpd/ft
C     T2         Aquifer transmissivity in J-direction between
C                I,J and I+1,J in gpd/ft
C     TRANS      Average aquifer transmissivity in gpd/ft
C     WBE        Water balance tolerance error as a percentage
C     ZDM        Partial penetration mine factor as a decimal
```

```
C       PP,PM,HH         Default values
C       PPA,SS2,SS1      Default values
C       PPC,PMC,SS3      Default values
C       AA,BB,CC,DD      Coefficients in water balance equations
      END
```

```
C
      SUBROUTINE INFO2(MAL,MAU,MAS,MDY,NL,LC1)
C
C     IN THIS SUBROUTINE, AQUIFER LAYERING, AQUITARD
C     STORATIVITY, DELAYED GRAVITY YIELD, AND NUMBER OF
C     AQUIFER LAYERS OR AQUIFER-AQUITARD UNITS ARE
C     USER SPECIFIED
C
C     DEFINE OUTPUT DEVISE
      OPEN(5,FILE='LPT1:')
C
   10 WRITE(*,20)
   20 FORMAT('0','Enter 1 to simulate aquifer layering'/1X,
     *'or 2 for no aquifer layering  '\)
      READ(*,*)MAL
      IF(MAL.NE.1.AND.MAL.NE.2)GOTO 10
   30 IF(MAL.EQ.1)THEN
      MAU=0
   32 WRITE(*,40)
   40 FORMAT('0','Enter 1 to simulate aquifer layering of'/1X,
     *'units 1 and 2, 2 to simulate aquifer layering of'/1X,
     *'units 3 and 4, or 3 to simulate aquifer layering of'/1X,
     *'units 1-4  '\)
      READ(*,*)MAU
      IF(MAU.NE.1.AND.MAU.NE.2.AND.MAU.NE.3)GOTO 32
      ENDIF
   45 WRITE(*,47)
   47 FORMAT('0','Enter 1 to simulate aquitard storativity or'/1X,
     *'delayed gravity yield under water table conditions or'/1X,
     *'2 to continue program  '\)
      READ(*,*)LC3
      IF(LC3.NE.1.AND.LC3.NE.2)GOTO 45
      IF(LC3.EQ.2)GOTO 90
   50 WRITE(*,60)
   60 FORMAT('0','Enter 1 to simulate aquitard storativity.'/1X,
     *'Simulation is active when time after pumping started in'/1X,
     *'days is less than 1.0 x aquitard storativity x aquitard'/1X,
     *'thickness in ft divided by aquitard hydraulic conduct-'/1X,
     *'ivity in gpd/sq ft and each pumping rate period exceeds'/1X,
     *'this time in the case of multi-pumping rate schedules. The'/1X,
     *'impacts of aquitard storativity are negligible after this'/1X,
     *'time. Enter 2 for no aquitard storativity simulation.  '\)
      READ(*,*)MAS
      IF(MAS.NE.1.AND.MAS.NE.2)GOTO 50
      IF(MAS.EQ.1)GOTO 92
   70 WRITE(*,80)
```

```
 80 FORMAT('0','Enter 1 to simulate delayed gravity yield'/1X,
   *'under water table conditions or 2 for no delayed gravity'/1X,
   *'yield simulation. The model for delayed gravity yield'/1X,
   *'assumes a single aquifer unit with time variable stor-'/1X,
   *'ativity and nodal recharge from gravity drainage. Delayed'/1X,
   *'gravity yield impacts are negligible when the elapsed time'/1X,
   *'in minutes is greater than 54000 x aquifer thickness in ft'/1X,
   *'x aquifer specific yield divided by aquifer vertical'/1X,
   *'hydraulic conductivity in gpd/sq ft  '\)
    READ(*,*)MDY
    IF(MDY.NE.1.AND.MDY.NE.2)GOTO 70
    GOTO 94
 90 MAS=2
 92 MDY=2
 94 IF(LC1.EQ.1)WRITE(*,100)
100 FORMAT('0','Aquifer units or layers may be simulated.'/1X,
   *'For aquifer layering, there must be either 2 or 4 units.'/1X,
   *'An aquifer layering unit consists of an aquifer layer and'/1X,
   *'associated inter-aquifer flow transfer. An aquifer-aquitard'/1X,
   *'unit consists of an aquifer and overlying aquitard. Number'/1X,
   *'of units must be less than 5. It is assumed that an'/1X,
   *'aquiclude underlies the lowermost aquifer or layer.'/1X,
   *'Units are numbered in order from top to bottom. Leaky'/1X,
   *'artesian conditions with a constant sourcebed head may')
    IF(LC1.EQ.1)WRITE(*,110)
110 FORMAT(1X,
   *'be simulated in the uppermost unit by specifying the'/1X,
   *'sourcebed head and aquitard hydraulic conductivity and'/1X,
   *'thickness. Nonleaky artesian or water table conditions are'/1X,
   *'simulated in the uppermost unit by setting aquitard'/1X,
   *'hydraulic conductivity and thickness to zero thus creating'/1X,
   *'an aquiclude above aquifer 1. By default, aquitard storat-'/1X,
   *'ivity is assumed to be negligible. If aquifer layering'/1X,
   *'is simulated then it is assumed that water table conditions'/1X,
   *'prevail in the uppermost layer at the top of the aquifer'/1X,
   *'system.  If delayed gravity yield is simulated then'/1X,
   *'the number of units or layers must be 1.')
    WRITE(*,116)
116 FORMAT('0','Enter integer number of units or layers  '\)
    READ(*,*)NL
    WRITE(5,120)NL
120 FORMAT(1X,'Number of units or layers=',I3)
    IF(LC1.EQ.2)GOTO 150
130 WRITE(*,140)
140 FORMAT('0','An infinite aquifer system may be'/1X,
   *'simulated by selecting grid spacings so that'/1X,
   *'water level decline is negligible along the'/1X,
   *'grid borders. A barrier boundary may be simulated'/1X,
   *'by assigning zero transmissivities and very small'/1X,
   *'storativities or storage factors outside the boundary of'/1X,
   *'interest. A constant head boundary(recharge boundary)'/1X,
   *'may be simulated by setting storativities of the nodes'/1X,
```

```
     *'along the position of the boundary to large values.'/1X,
     *'A specified head boundary may be simulated by setting'/1X,
     *'heads at the nodes along the position of the boundary'/1X,
     *'to a selected value. A specified flow boundary may be'/1X,
     *'simulated by assigning modified transmissivities and'/1X,
     *'and storativities along the boundary of interest. The'/1X,
     *'precision of water balance calculations at constant'/1X,
     *'head nodes may not be high because although head loss'/1X,
     *'is small, storativity is high.')
      PAUSE
  150 CLOSE (5)
      END
```

```
C
      SUBROUTINE MINE(TIME,ISTEP,DELTA,DELX,DELY,THICK,
     *SF1,SF2,S1,S2,H,HO,T1,T2,P1,P2,QMD,BOT,MDM1,NDMN,MDM2,
     *MDM3,MDM4,LC1)
C
C     THIS SUBROUTINE ENTERS AND DISPLAYS MINE
C     LOCATION AND GEOMETRY DATA BASE
C
C     METACOMMANDS
C
$LARGE
C
C     DECLARATIONS
C
      DIMENSION TIME(50),DELTA(50),DELX(21),DELY(21),THICK(20,20,4),
     *SF1(20,20,4),SF2(20,20,4),T1(20,20,4),T2(20,20,4),P1(20,20,4),
     *P2(20,20,4),BOT(20,20,4),H(20,20,4),HO(20,20,4),QMD(20,20,4),
     *S1(20,20,4),S2(20,20,4)
C
    5 WRITE(*,7)
    7 FORMAT('0','Enter 1 to simulate mines'/1X,
     *'or 2 for no mines  '\)
      READ(*,*)MDM1
      IF(MDM1.NE.1.AND.MDM1.NE.2)GOTO 5
      IF(MDM1.EQ.1)THEN
      OPEN(5,FILE='LPT1:')
    8 IF(LC1.EQ.2)GOTO 25
      WRITE(*,10)
   10 FORMAT('0','Mines are simulated by setting'/1X,
     *'storage factors at 1E21 to create constant heads'/1X,
     *'at selected nodes. Mine advancement is simul-'/1X,
     *'ated by setting constant heads sequentially in time'/1X,
     *'at adjacent nodes. Grid spacing at and near mine'/1X,
     *'nodes should be square and taylored to mine width'/1X,
     *'and length which should be equal. Transmissivities'/1X,
     *'at and near mine nodes are adjusted with a mine'/1X,
     *'simulation factor so that mine inflow faces are'/1X,
     *'located between grid lines at distances one half'/1X,
     *'grid spacings,thus,taking into account the'/1X,
     *'mine cavity centered at the mine node.')
      PAUSE
      WRITE(*,20)
   20 FORMAT('0','Mine drifts,rooms,pillars,longholes,haulage'/1X,
     *'ways,stopes,etc. are assumed to be so effective in'/1X,
     *'aquifer drainage that the outer most mine-aquifer face'/1X,
     *'represents the actual mine inflow face. At least 2'/1X,
```

```
     *'nodes should separate mines and aquifer boundaries.'/1X,
     *'The mine simulation factor may consider degree of'/1X,
     *'mine penetration into the aquifer and the ratio of'/1X,
     *'horizontal to vertical aquifer hydraulic conductivity.'/1X,
     *'Heads are maintained above the aquifer base.')
      PAUSE
   25 WRITE(*,50)
   50 FORMAT('0','Enter integer number of mine nodes to be'/1X,
     *'considered during this time increment  '\)
      READ(*,*)NDMN
      WRITE(5,60)NDMN
   60 FORMAT('0','Number of mine nodes to be considered'/1X,
     *'during this time increment=',I3)
      DO 440 N=1,NDMN
      WRITE(*,80)N
   80 FORMAT(1X,'Mine node number',I3)
      WRITE(5,90)N
   90 FORMAT(1X,'Mine node number',I3)
      WRITE(*,92)
   92 FORMAT(1X,'Enter integer aquifer unit or layer'/1X,
     *'in which mine is located  '\)
      READ(*,*)K
      WRITE(5,94)K
   94 FORMAT(1X,'Aquifer unit or layer in which'/1X,
     *'mine is located=',I3)
      WRITE(*,100)
  100 FORMAT(1X,'Enter integer I-coordinate of'/1X,
     *'mine node  '\)
      READ(*,*)I
      WRITE(5,110)I
  110 FORMAT(1X,'I-coordinate of mine node ',I3)
      WRITE(*,120)
  120 FORMAT(1X,'Enter integer J-coordinate of'/1X,
     *'mine node  '\)
      READ(*,*)J
      WRITE(5,130)J
  130 FORMAT(1X,'J-coordinate of mine node ',I3)
      WRITE(*,140)
  140 FORMAT(1X,'Enter real number mine floor'/1X,
     *'elevation in ft. Must be 1 ft above'/1X,
     *'aquifer base elevation  '\)
      READ(*,*)EMF
      IF(EMF.LE.BOT(I,J,K))EMF=BOT(I,J,K)+0.1
      WRITE(5,150)EMF
  150 FORMAT(1X,'Mine floor elevation in ft=',F9.2)
  160 WRITE(*,170)
  170 FORMAT('0','Enter 1 if mine node is active, 2'/1X,
     *'if mine node is inactive, or 3 if mine node is'/1X,
     *'being backfilled during this time increment. Mine'/1X,
     *'node must first be active before it can be'/1X,
     *'inactive or backfilled  '\)
      READ(*,*)MDM2
```

```
      IF(MDM2.NE.1.AND.MDM2.NE.2.AND.MDM2.NE.3)GOTO 160
      IF(MDM2.EQ.1)WRITE(5,180)
  180 FORMAT(1X,'Mine node is active')
      IF(MDM2.EQ.2)WRITE(5,190)
  190 FORMAT(1X,'Mine node is inactive')
      IF(MDM2.EQ.3)WRITE(5,200)
  200 FORMAT(1X,'Mine node is being backfilled')
  210 WRITE(*,220)
  220 FORMAT('0','Enter 1 if mine node first became active'/1X,
     *'during this time increment and you want to include'/1X,
     *'mine cavity drainage in mine discharge rate or 2 to'/1X,
     *'continue program   '\)
      READ(*,*)MDM3
      IF(MDM3.NE.1.AND.MDM3.NE.2)GOTO 210
  230 WRITE(*,240)
  240 FORMAT('0','Enter 1 if mine at node fully penetrates'/1X,
     *'the aquifer or 2 if mine at node partially'/1X,
     *'penetrates the aquifer   '\)
      READ(*,*)MDM4
      IF(MDM4.NE.1.AND.MDM4.NE.2)GOTO 230
      IF(MDM4.EQ.1)WRITE(5,250)
  250 FORMAT(1X,'Mine at node fully penetrates the aquifer')
      IF(MDM4.EQ.2)WRITE(5,260)
  260 FORMAT(1X,'Mine at node partially penetrates the aquifer')
      IF(MDM2.EQ.1)THEN
C
C     CREATE CONSTANT HEAD NODES
C
      SF1(I,J,K)=1E21
      SF2(I,J,K)=1E21
      H(I,J,K)=EMF
      HO(I,J,K)=EMF
      ENDIF
      IF(MDM2.EQ.1.AND.MDM4.EQ.1)THEN
C
C     ADJUST TRANSMISSIVITY FOR MINE CAVITY
C
      T1(I,J,K)=T1(I,J,K)/0.5
      T1(I,J-1,K)=T1(I,J-1,K)/0.5
      T2(I,J,K)=T2(I,J,K)/0.5
      T2(I-1,J,K)=T2(I-1,J,K)/0.5
      ENDIF
      IF(MDM4.EQ.1)GOTO 272
      WRITE(*,270)
  270 FORMAT(1X,'Enter real number partial penetration'/1X,
     *'mine factor determined with a cross-sectional'/1X,
     *'finite-difference model by comparing mine flows'/1X,
     *'with full and partial penetration (usually ranges'/1X,
     *'from 0.51 to 0.81   '\)
      READ(*,*)ZDM
      IF(MDM2.EQ.1.AND.MDM4.EQ.2)THEN
      T1(I,J,K)=T1(I,J,K)/ZDM
```

```
      T1(I,J-1,K)=T1(I,J-1,K)/ZDM
      T2(I,J,K)=T2(I,J,K)/ZDM
      T2(I-1,J,K)=T2(I-1,J,K)/ZDM
      ENDIF
  272 IF(MDM2.EQ.2)THEN
C
C     CALCULATE STORATIVITY FACTOR WITHOUT CONSTANT HEAD
C     AND READJUST TRANSMISSIVITY
C
      SF1(I,J,K)=S1(I,J,K)*7.48*DELX(I)*DELY(J)
      SF2(I,J,K)=S2(I,J,K)*7.48*DELX(I)*DELY(J)
      ENDIF
      IF(MDM2.EQ.2.AND.MDM4.EQ.1)THEN
      T1(I,J,K)=T1(I,J,K)*0.5
      T1(I,J-1,K)=T1(I,J-1,K)*0.5
      T2(I,J,K)=T2(I,J,K)*0.5
      T2(I-1,J,K)=T2(I-1,J,K)*0.5
      ENDIF
      IF(MDM2.EQ.2.AND.MDM4.EQ.2)THEN
      T1(I,J,K)=T1(I,J,K)*ZDM
      T1(I,J-1,K)=T1(I,J-1,K)*ZDM
      T2(I,J,K)=T2(I,J,K)*ZDM
      T2(I-1,J,K)=T2(I-1,J,K)*ZDM
      ENDIF
      IF(MDM2.EQ.3)THEN
      SF1(I,J,K)=S1(I,J,K)*7.48*DELX(I)*DELY(J)
      SF2(I,J,K)=S2(I,J,K)*7.48*DELX(I)*DELY(J)
      ENDIF
      IF(MDM2.EQ.3.AND.MDM4.EQ.1)THEN
      T1(I,J,K)=T1(I,J,K)*0.5
      T1(I,J-1,K)=T1(I,J-1,K)*0.5
      T2(I,J,K)=T2(I,J,K)*0.5
      T2(I-1,J,K)=T2(I-1,J,K)*0.5
      ENDIF
      IF(MDM2.EQ.3.AND.MDM4.EQ.2)THEN
      T1(I,J,K)=T1(I,J,K)*ZDM
      T1(I,J-1,K)=T1(I,J-1,K)*ZDM
      T2(I,J,K)=T2(I,J,K)*ZDM
      T2(I-1,J,K)=T2(I-1,J,K)*ZDM
      ENDIF
      IF(MDM2.EQ.3.AND.MDM4.EQ.1)THEN
      WRITE(*,280)
  280 FORMAT(1X,'Enter real number hydraulic conductivity'/1X,
     *'of backfill materials in gpd/sq ft  '\)
      READ(*,*)HCBM
      WRITE(5,290)HCBM
  290 FORMAT(1X,'Hydraulic conductivity of backfill materials'/1X,
     *'in gpd/sq ft–',F9.2)
C
C     ADJUSTMENTS FOR VARIABLE GRID
C
      P1(I,J,K)=HCBM*2*DELX(I)/(DELY(J)+DELY(J+1))
```

```
      P2(I,J,K)=HCBM*2*DELY(J)/(DELX(I)+DELX(I+1))
      T1(I,J,K)=P1(I,J,K)*THICK(I,J,K)
      T2(I,J,K)=P2(I,J,K)*THICK(I,J,K)
      WRITE(*,300)
300 FORMAT(1X,'Enter real number storativity of backfill'/1X,
   *'materials as a decimal  '\)
      READ(*,*)STOR1
      S1(I,J,K)=STOR1
      WRITE(5,310)STOR1
310 FORMAT(1X,'Storativity of backfill materials'/1X,
   *'as a decimal=',F10.8)
      WRITE(*,320)
320 FORMAT(1X,'Enter specific yield of backfill'/1X,
   *'materials as a decimal  '\)
      READ(*,*)STOR2
      WRITE(5,330)STOR2
330 FORMAT(1X,'Specific yield of backfill materials'/1X,
   *'as a decimal=',F10.8)
      S2(I,J,K)=STOR2
      SF1(I,J,K)=S1(I,J,K)*7.48*DELX(I)*DELY(J)
      SF2(I,J,K)=S2(I,J,K)*7.48*DELX(I)*DELY(J)
      ENDIF
      IF(MDM2.EQ.3.AND.MDM4.EQ.2)THEN
      WRITE(*,340)
340 FORMAT(1X,'Enter real number hydraulic conductivity'/1X,
   *'of backfill materials in gpd/sq ft  '\)
      READ(*,*)HCBM
      WRITE(5,350)HCBM
350 FORMAT(1X,'Hydraulic conductivity of backfill'/1X,
   *'materials in gpd/sq ft=',F9.2)
      WRITE(*,360)
360 FORMAT(1X,'Enter real number artesian storativity'/1X,
   *'of backfill materials as a decimal  '\)
      READ(*,*)STOR1
      WRITE(5,380)STOR1
380 FORMAT(1X,'Artesian storativity of backfill'/1X,
   *'materials as a decimal=',F10.8)
      WRITE(*,400)
400 FORMAT(1X,'Enter real number specific yield of'/1X,
   *'backfill materials as a decimal  '\)
      READ(*,*)STOR2
      WRITE(5,410)STOR2
410 FORMAT(1X,'Specific yield of backfill materials'/1X,
   *'as a decimal=',F10.8)
      WRITE(*,420)
420 FORMAT(1X,'Enter real number depth of penetration'/1X,
   *'of mine below aquifer top in ft  '\)
      READ(*,*)DPEN
      WRITE(5,430)DPEN
430 FORMAT(1X,'Depth of penetration of mine below'/1X,
   *'aquifer top in ft=',F8.2)
      T1B=HCBM*DPEN
```

```
      T2B=HCBM*DPEN
      T1A=P1(I,J,K)*(THICK(I,J,K)-DPEN)
      T2A=P2(I,J,K)*(THICK(I,J,K)-DPEN)
      T1(I,J,K)=T1A+T1B
      T2(I,J,K)=T2A+T2B
      T1(I,J,K)=T1(I,J,K)*2*DELX(I)/(DELY(J)+DELY(J+1))
      T2(I,J,K)=T2(I,J,K)*2*DELY(J)/(DELX(I)+DELX(I+1))
      P1(I,J,K)=T1(I,J,K)/THICK(I,J,K)
      P2(I,J,K)=T2(I,J,K)/THICK(I,J,K)
      S1(I,J,K)=STOR1
      S2(I,J,K)=STOR2
      SF1(I,J,K)=S1(I,J,K)*7.48*DELX(I)*DELY(J)
      SF2(I,J,K)=S2(I,J,K)*7.48*DELX(I)*DELY(J)
      ENDIF
      IF(MDM2.EQ.1.AND.MDM3.EQ.1)THEN
      IF(MDM4.EQ.1)THEN
      QMD(I,J,K)=QMD(I,J,K)+(7.48*DELX(I)*DELY(J)*THICK(I,J,K)
     */DELTA(ISTEP))
      ENDIF
      IF(MDM2.EQ.1.AND.MDM4.EQ.2)THEN
      QMD(I,J,K)=QMD(I,J,K)+(7.48*DELX(I)*DELY(J)*DPEN
     */DELTA(ISTEP))
      ENDIF
      ENDIF
  440 CONTINUE
  442 WRITE(*,443)
  443 FORMAT('0','Enter 1 to revise mine data base'/1X,
     *'or 2 to continue program  '\)
      READ(*,*)MDM5
      IF(MDM5.NE.1.AND.MDM5.NE.2)GOTO 442
      IF(MDM5.EQ.1)GOTO 8
      CLOSE (5)
      ENDIF
  445 END
```

```
C
      SUBROUTINE PARPEN
C
C      THIS SUBROUTINE CALCULATES WELL PARTIAL
C      PENETRATION IMPACTS
C
C      METACOMMANDS
C
$LARGE
C
C      THIS SUBROUTINE APPLIES TO AQUIFER UNITS BUT
C      NOT TO AQUIFER LAYERING NOR MULTI-UNIT OR LAYER
C      PRODUCTION WELLS
C
   10 WRITE(*,20)
   20 FORMAT('0','Enter 1 to simulate well partial'/1X,
     *'penetration conditions in units or 2 to simulate full'/1X,
     *'penetration conditions. Enter 2 if aquifer layering'/1X,
     *'is simulated. The distance from the production'/1X,
     *'well beyond which partial penetration impacts are'/1X,
     *'negligible in ft is equal to 1.5 x aquifer thickness'/1X,
     *'in ft x square root of the ratio of aquifer horizontal'/1X,
     *'and vertical hydraulic conductivity. Please note that'/1X,
     *'multi-unit or layer production wells are assumed to be'/1X,
     *'fully penetrating  '\)
      READ(*,*)MPP
      IF(MPP.NE.1.AND.MPP.NE.2)GOTO 10
      IF(MPP.EQ.2)GOTO 490
C
C      DATA BASE SPECIFICATION
C
   25 WRITE(*,30)
   30 FORMAT('0','The following data base should be based on'/1X,
     *'previous calculations or other available information'/)
   35 WRITE(*,40)
   40 FORMAT(1X,'Enter real number production well radius'/1X,
     *'in ft  '\)
      READ(*,*)RW
      WRITE(*,50)
   50 FORMAT(1X,'Enter real number production well discharge'/1X,
     *'rate in gpm  '\)
      READ(*,*)Q
   55 WRITE(*,60)
   60 FORMAT('0','Enter 1 for calculation of partial penetration'/1X,
     *'impacts in an observation well or 2 for calculation of'/1X,
     *'partial penetration impacts in a production well  '\)
```

```
      READ(*,*)MPO
      IF(MPO.NE.1.AND.MPO.NE.2)GOTO 55
      IF(MPO.EQ.2)GOTO 80
      WRITE(*,70)
   70 FORMAT('0','Enter real number distance from production'/1X,
     *'well to observation well in ft  '\)
      READ(*,*)RWO
      RW=RWO
   80 WRITE(*,90)
   90 FORMAT('0','Enter real number average aquifer transmissivity'/1X,
     *'of model in gpd/ft  '\)
      READ(*,*)TRANS
      WRITE(*,100)
  100 FORMAT(1X,'Enter real number average aquifer storativity'/1X,
     *'of model as a decimal  '\)
      READ(*,*)STOR
      WRITE(*,110)
  110 FORMAT(1X,'Enter real number time after pumping started'/1X,
     *'in minutes  '\)
      READ(*,*)TIME
      WRITE(*,120)
  120 FORMAT(1X,'Enter real number ratio of aquifer vertical'/1X,
     *'hydraulic conductivity to aquifer horizontal hydraulic'/1X,
     *'conductivity as a decimal   '\)
      READ(*,*)RATIO
      WRITE(*,130)
  130 FORMAT(1X,'Enter real number aquifer thickness in ft  '\)
      READ(*,*)THICK
      WRITE(*,140)
  140 FORMAT(1X,'Enter real number distance from aquifer top'/1X,
     *'to bottom of production well in ft  '\)
      READ(*,*)DLP
      WRITE(*,150)
  150 FORMAT(1X,'Enter real number distance from aquifer top'/1X,
     *'to top of production well screen in ft  '\)
      READ(*,*)DDP
      IF(MPO.EQ.2) THEN
      DLO=(DLP+DDP)/2
      DDO=DLP-1
      GOTO 175
      ENDIF
      WRITE(*,160)
  160 FORMAT(1X,'Enter real number distance from aquifer top'/1X,
     *'to observation well bottom in ft  '\)
      READ(*,*)DLO
      WRITE(*,170)
  170 FORMAT(1X,'Enter real number distance from aquifer top'/1X,
     *'to top of observation well screen in ft  '\)
      READ(*,*)DDO
  175 WRITE(*,180)
  180 FORMAT(1X,'Enter real number drawdown in production'/1X,
     *'or observation well with full penetration based on'/1X,
```

```
     *'previous calculations   '\)
      READ(*,*)DR
      U=2693*RW**2*STOR/(TRANS*TIME)
      Z=0
C
C     CALCULATE PARTIAL PENETRATION WELL FUNCTION
C     USING POLYNOMIAL APPROXIMATIONS
C
      DO 420 K=1,50
      S=K*3.1416*RW*SQRT(RATIO)/THICK
      IF(U.GT.5.0)THEN
      WS=0.0
      GOTO 400
      ENDIF
      IF(S.GT.2.0)THEN
      SM=-((S-2*U)/(2*SQRT(U)))
      IF(SM.LT.0.0)THEN
      SM=ABS(SM)
      SP=1+.0705230784*SM+.0422820123*SM**2+9.2705272E-3*SM**3+
     *1.520143E-4*SM**4+2.765672E-4*SM**5+4.30638E-5*SM**6
      IF(SP.GT.100)THEN
      SN=0.0
      GOTO 250
      ENDIF
      SN=1/SP**16
  250 WS=(3.1416/SQRT(2*S))*EXP(-S)*(2-SN)
      GOTO 400
      ENDIF
      SP=1+.0705230784*SM+.0422820123*SM**2+9.2705272E-3*SM**3+
     *1.520143E-4*SM**4+2.765672E-4*SM**5+4.30638E-5*SM**6
      IF(SP.GT.100)THEN
      SN=0.0
      GOTO 270
      ENDIF
      SN=1/SP**16
  270 WS=(3.1416/SQRT(2*S))*EXP(-S)*SN
      GOTO 400
      ENDIF
  300 IF(U.GE.1.0)THEN
      WU=U**4+8.5733287401*U**3+18.0590169730*U**2+8.6347608925*U+
     *.2677737343
      WUP=U**4+9.5733223454*U**3+25.6329561486*U**2+
     *21.0996530827*U+3.9584969228
      WU=WU/WUP
      WU=WU/(U*EXP(U))
      GOTO 320
      ENDIF
      WU=-ALOG(U)-.57721566+.99999193*U-.24991055*U**2+
     *.05519968*U**3-9.76004E-3*U**4+1.07857E-3*U**5
  320 SL=S/3.75
      V=1+3.5156229*SL**2+3.0899424*SL**4+1.2067492*SL**6+
     *.2659732*SL**8+.0360768*SL**10+.0045813*SL**12
```

```
      F=S/2
      H=-ALOG(F)*V-.57721566+.4227840*F**2+.23069756*F**4+
     *.0348859*F**6+2.62698E-3*F**8+.00010750*F**10+
     *.00000740*F**12
      IF(S.EQ.0.0)THEN
      WS=WU
      GOTO 400
      ENDIF
      SN=S**2/(4*U)
      IF(SN.GT.5.0)THEN
      WS=2*H
      GOTO 400
      ENDIF
      IF(U.LE.0.9)GOTO 340
      A=U+.5858
      B=U+3.414
      C=S**2/4
      WS=1.5637*EXP(-A-C/A)/A+4.54*EXP(-B-C/B)/B
      GOTO 400
  340 IF(U.LT.0.5)GOTO 360
      IF(U.GT.S/2)GOTO 350
      C=-(1.75*U)**(.448*S)
      WS=2*H-4.8*10**C
      GOTO 400
  350 WS=WU-(S/(4.7*U**6))**2
      GOTO 400
  360 IF(U.GT.0.01.AND.S.LT.0.1)GOTO 350
      IF(SN.LT.1.0)THEN
      WN=-ALOG(SN)+.99999193*SN-.57721566-.24991055*SN**2+
     *.05519968*SN**3-9.76004E-3*SN**4+1.07857E-3*SN**5
      WS=2*H-WN*V
      GOTO 400
      ENDIF
      WN=SN**4+8.57332874*SN**3+18.0590169730*SN**2+8.6347608925*SN+
     *.2677737343
      WNP=SN**4+9.5733223454*SN**3+25.6329561486*SN**2+
     *21.0996530827*SN+3.9584969228
      WN=WN/WNP
      WN=WN/(SN*EXP(SN))
      WS=2*H-WN*V
  400 AA=SIN(K*3.1416*(DLP/THICK))-SIN(K*3.1416*(DDP/THICK))
      BB=SIN(K*3.1416*(DLO/THICK))-SIN(K*3.1416*(DDO/THICK))
      Z=Z+1/K**2*(AA*BB*WS)
  420 CONTINUE
      WUPAR=2*THICK**2*Z/(3.1416**2*(DLP-DDP)*(DLO-DDO))
      DPP=114.6*Q*WUPAR/TRANS
      WRITE(*,450)DPP
  450 FORMAT('0','Drawdown due to well partial penetration'/1X,
     *'in ft=',F9.3)
      DTOT=DR+DPP
      WRITE(*,460)DTOT
  460 FORMAT(1X,'Total drawdown with partial penetration'/1X,
```

```
   *'in ft=',F9.3)
470 WRITE(*,480)
480 FORMAT('0','Enter 1 for another partial penetration'/1X,
   *'impact calculation or 2 to continue program  '\)
    READ(*,*)MP3
    IF(MP3.NE.1.AND.MP3.NE.2)GOTO 470
    IF(MP3.EQ.1)GOTO 25
490 END
```

```
C
      SUBROUTINE PROFIL(H,NL)
C
C      METACOMMANDS
C
$LARGE
C
C      THIS SUBROUTINE CONVERTS PLAN VIEW AQUIFER
C      UNIT OR LAYER HEAD DATA BASE INTO HEAD
C      PROFIL DATA BASE
C
C      DECLARATIONS
C
      CHARACTER*14FNAME
      DIMENSION H(20,20,4),INODE(20),JNODE(20),HNODE(50,4)
C
C      DEFINE OUTPUT DEVISE
C
      OPEN(5,FILE='LPT1:')
C
   30 WRITE(*,40)
   40 FORMAT('0','Enter 1 to convert plan view aquifer'/1X,
     *'unit or layer head data base into a head profile'/1X,
     *'data base or 2 for no profile  '\)
      READ(*,*)MP1
      IF(MP1.NE.1.AND.MP1.NE.2)GOTO 30
      IF(MP1.EQ.2)GOTO 810
      WRITE(*,130)
  130 FORMAT('0','Enter integer number of nodes on'/1X,
     *'profile (must be < 51)  '\)
      READ(*,*)NODE
      WRITE(5,140)NODE
  140 FORMAT('0','Number of nodes on profile=',I3)
      DO 390 L=1,NODE
      WRITE(*,380)L
  380 FORMAT(1X,'Node number=',I3)
      WRITE(*,382)
  382 FORMAT(1X,'Enter integer I-coordinate of node'/1X,
     *'proceeding from left to right on profile  '\)
      READ(*,*)INODE(L)
      WRITE(5,384)L,INODE(L)
  384 FORMAT(1X,'I-coordinate of node',I3,' =', I3)
      WRITE(*,385)
  385 FORMAT(1X,'Enter integer J-coordinate of node  '\)
      READ(*,*)JNODE(L)
      WRITE(5,387)L,JNODE(L)
```

```
387 FORMAT(1X,'J-coordinate of node',I3,' =',I3)
390 CONTINUE
    DO 450 L=1,NODE
    DO 440 K=1,NL
    I=INODE(L)
    J=JNODE(L)
    HNODE(L,K)=H(I,J,K)
440 CONTINUE
450 CONTINUE
460 WRITE(*,470)
470 FORMAT('0','Enter 1 for screen display of head profile'/1X,
   *'or 2 to continue program  '\)
    READ(*,*)N1C
    IF(N1C.NE.1.AND.N1C.NE.2)GOTO 460
    IF(N1C.EQ.1)THEN
    WRITE(*,472)
472 FORMAT('0','Use Ctrl-Num Lock keys to'/1X,
   *'control screen scrolling')
    PAUSE
    WRITE(*,480)
480 FORMAT('0','HEAD IN AQUIFER UNITS OR AQUIFER'/1X,
   *'LAYERS IN FT:')
500 WRITE(*,510)
510 FORMAT('0','UNIT OR                    NODE NUMBER'/1X,
   *'LAYER NO.'/)
    N1=1
    N2=8
    IF(NODE.LT.N2)N2=NODE
610 WRITE(*,620)(I,I=N1,N2)
620 FORMAT(3X,8I9)
    DO 640 J=1,NL
    WRITE(*,622)J,(HNODE(I,J),I=N1,N2)
622 FORMAT(I3,8F9.2)
640 CONTINUE
    IF(N2.EQ.NODE)GOTO 650
    N1=N1+8
    N2=N2+8
    IF(NODE.LT.N2)N2=NODE
    GOTO 610
650 ENDIF
660 WRITE(*,665)
665 FORMAT('0','Enter 1 for hard copy of head profile'/1X,
   *'or 2 to continue program  '\)
    READ(*,*)N2C
    IF(N2C.NE.1.AND.N2C.NE.2)GOTO 660
    IF(N2C.EQ.1)THEN
    N1=1
    N2=8
    IF(NODE.LT.N2)N2=NODE
    WRITE(5,680)
680 FORMAT('0','HEAD IN AQUIFER UNITS OR AQUIFER'/1X,
   *'LAYERS IN FT:')
```

```
690 WRITE(5,700)
700 FORMAT('0','UNIT OR                    NODE NUMBER'/1X,
   *'LAYER NO.'/)
710 WRITE(5,720)(I,I=N1,N2)
720 FORMAT(3X,8I9)
    DO 740 J=1,NL
    WRITE(5,722)J,(HNODE(I,J),I=N1,N2)
722 FORMAT(I3,8F9.2)
740 CONTINUE
    IF(N2.EQ.NODE)GOTO 750
    N1=N1+8
    N2=N2+8
    IF(NODE.LT.N2)N2=NODE
    GOTO 710
750 ENDIF
    CLOSE (5)
810 END
```

```
C
      SUBROUTINE RESLT(DELX,DELY,T1,T2,H,Q,TIME,ISTEP,
     *R,IP,JP,KP,RS,MAS,HO,SF1,DELTA,SF2,MDY,QDY,MAP1,A7,A8,
     *QMA,KUM,KLM,IMP,JMP,QMD,QRD,QST,QDRAIN,QF,MF1,
     *MDM1,MS1,MD1,
     *MUS,HSB,CH,NL,MDM2)
C
C     THIS SUBROUTINE DISPLAYS SELECTED COMPUTATION RESULTS
C
C     METACOMMANDS
C
$LARGE
C
C     DECLARATIONS
C
      DIMENSION DELX(21),DELY(21),T1(20,20,4),T2(20,20,4),H(20,20,4),
     *Q(20,20,4),TIME(50),IP(50),JP(50),KP(4),R(20,20,4),
     *RS(20,20,4),HO(20,20,4),DELTA(50),SF1(20,20,4),SF2(20,20,4),
     *QDY(20,20,1),QMD(20,20,4),QRD(20,20,4),QST(20,20,1),
     *QDRAIN(20,20,1),QF(20,20,4),CH(20,20,4)
C
      OPEN(5,FILE='LPT1:')
      MDM6=0
  200 WRITE(*,210)
  210 FORMAT('0','Enter 1 for display of production well head'/1X,
     *'and well loss, flowing well discharge rate, mine'/1X,
     *'drainage rate, multi-aquifer production well head,'/1X,
     *'river induced infiltration rate, drain discharge'/1X,
     *'rate, or water balance data. Enter 2 to continue'/1X,
     *'program  '\)
      READ(*,*)ND1
      IF(ND1.NE.1.AND.ND1.NE.2)GOTO 200
      IF(ND1.EQ.2)GOTO 790
  260 WRITE(*,270)
  270 FORMAT('0','Enter 1 for production well heads'/1X,
     *'or 2 to continue program. If well partial penetration'/1X,
     *'or well storage capacity are simulated then enter 1. If'/1X,
     *'only multi-unit or layer production well is simulated then'/1X,
     *'enter 2  '\)
      READ(*,*)NRH
      IF(NRH.NE.1.AND.NRH.NE.2)GOTO 260
      IF(NRH.EQ.1)THEN
  280 WRITE(*,290)
  290 FORMAT('0','Enter integer production well number  '\)
      READ(*,*)L
      WRITE(*,300)
```

```
  300 FORMAT(1X,'Enter real number effective radius'/1X,
     *'of production well in ft  '\)
      READ(*,*)RAD
      I=IP(L)
      J=JP(L)
      K=KP(L)
C     SOLVE THIEM EQUATION WITH DRAWDOWN AT RW AND AT
C     A POINT EQUAL TO THE EQUVALENT WELL BLOCK RADIUS
C
      DELA=0.28*((T1(I,J,K)/T2(I,J,K))**.5*DELX(I)**2+
     *(T2(I,J,K)/T1(I,J,K))**.5*DELY(J)**2)**.5/
     *((T1(I,J,K)/T2(I,J,K))**.25+(T2(I,J,K)/T1(I,J,K))**.25)
      DEL1=DELA/RAD
      DEL2=ALOG(DEL1)
      TRAN=(T1(I,J,K)*T2(I,J,K))**.5
      HP=H(I,J,K)-(229*(Q(I,J,K)/(1440*TRAN))*DEL2)
C
C     SIMULATE WELL LOSS
C
  302 WRITE(*,303)
  303 FORMAT('0','Enter 1 to simulate well loss or 2'/1X,
     *'for no well loss  '\)
      READ(*,*)LWL
      IF(LWL.NE.1.AND.LWL.NE.2)GOTO 302
      IF(LWL.EQ.1)THEN
      WRITE(*,304)
  304 FORMAT('0','Enter real number well loss coefficient in'/1X,
     *'sec**2/ft**5. Usually ranges from .1 to 100'/1X,
     *'sec**2/ft**5  '\)
      READ(*,*)WLC
      WRITE(5,306)WLC
  306 FORMAT(1X,'Well loss coefficient in sec**2/ft**5=',F7.2)
      QCFS=Q(I,J,K)/(1440*449)
      WLS=WLC*QCFS**2
      HP=HP-WLS
      ENDIF
      WRITE(*,310)HP
  310 FORMAT(1X,'Head or drawdown in ft=',F9.2)
      WRITE(5,320)L,HP
  320 FORMAT('0','Head or drawdown in production well',I3,/1X,
     *' in ft=',F9.2)
  330 WRITE(*,340)
  340 FORMAT('0','Enter 1 for another production well'/1X,
     *'or 2 to continue program  '\)
      READ(*,*)NAP
      IF(NAP.NE.1.AND.NAP.NE.2)GOTO 330
      IF(NAP.EQ.1)GOTO 280
      ENDIF
      IF(MF1.EQ.1)THEN
  341 WRITE(*,342)
  342 FORMAT('0','Enter 1 for discharge rate of flowing'/1X,
     *'well or 2 to continue program  '\)
```

```
      READ(*,*)MF3
      IF(MF3.NE.1.AND.MF3.NE.2)GOTO 341
      IF(MF3.EQ.2)GOTO 351
  343 WRITE(*,344)
  344 FORMAT('0','Enter integer I-coordinate for flowing'/1X,
     *'well whose discharge rate display is desired  '\)
      READ(*,*)I
      WRITE(*,345)
  345 FORMAT(1X,'Enter integer J-coordinate for flowing'/1X,
     *'well whose discharge rate display is desired  '\)
      READ(*,*)J
      WRITE(*,346)
  346 FORMAT(1X,'Enter integer aquifer unit or layer'/1X,
     *'number in which flowing well is open  '\)
      READ(*,*)K
      FRATE=QF(I,J,K)/1440
      WRITE(*,347)I,J,K,TIME(ISTEP),FRATE
  347 FORMAT(1X,'Discharge rate from flowing well node with'/1X,
     *'I-coordinate=',I3,' and J-coordinate=',I3,/1X,
     *' unit or layer no.=',I3,' at time',F10.3,' days is',F9.2,' gpm')
      WRITE(5,348)I,J,K,TIME(ISTEP),FRATE
  348 FORMAT(1X,'Discharge rate from flowing well node with'/1X,
     *'I-coordinate=',I3,' and J-coordinate=',I3,/1X,
     *' unit or layer no.=',I3,' at time',F10.3,' days is',F9.2,' gpm')
  349 WRITE(*,350)
  350 FORMAT('0','Enter 1 for another flowing'/1X,
     *'well discharge rate or 2 to continue program  '\)
      READ(*,*)MF4
      IF(MF4.NE.1.AND.MF4.NE.2)GOTO 349
      IF(MF4.EQ.1)GOTO 341
      ENDIF
  351 IF(MDM1.EQ.1.AND.MDM2.EQ.1)THEN
      IF(MDM6.EQ.0)THEN
      WRITE(*,352)
  352 FORMAT('0','Enter 1 for mine node drainage rate'/1X,
     *'display or 2 to continue program  '\)
      READ(*,*)MN6
      IF(MN6.NE.1.AND.MN6.NE.2)GOTO 351
      ENDIF
      IF(MN6.EQ.1)THEN
      WRITE(*,353)
  353 FORMAT('0','Enter integer I-coordinate for mine node'/1X,
     *'whose drainage rate display is desired  '\)
      READ(*,*)I
      WRITE(*,354)
  354 FORMAT(1X,'Enter integer J-coordinate for mine node'/1X,
     *'whose drainage rate display is desired  '\)
      READ(*,*)J
      WRITE(*,355)
  355 FORMAT(1X,'Enter integer aquifer unit or layer'/1X,
     *'number in which mine is located  '\)
      READ(*,*)K
```

```
      DRATE=(QMD(I,J,K)+QRD(I,J,K))/1440
      WRITE(*,356)I,J,K,TIME(ISTEP),DRATE
  356 FORMAT(1X,'Drainage/dewatering rate from mine node with'/1X,
     *'I-coordinate=',I3,' and J-coordinate=',I3,/1X,
     *' unit or layer no.=',I3,' at time',F10.3,' days is',F12.2,'gpm')
      WRITE(5,357)I,J,K,TIME(ISTEP),DRATE
  357 FORMAT(1X,'Drainage/dewatering rate from mine node with'/1X,
     *'I-coordinate=',I3,' and J-coordinate=',I3,/1X,
     *' unit or layer no.=',I3,' at time',F10.3,' days is',F12.2,'gpm')
  358 WRITE(*,359)
  359 FORMAT('0','Enter 1 for another mine node'/1X,
     *'drainage rate or 2 to continue program  '\)
      READ(*,*)MDM6
      IF(MDM6.NE.1.AND.MDM6.NE.2)GOTO 358
      IF(MDM6.EQ.1)GOTO 351
      ENDIF
      ENDIF
  440 WRITE(*,450)
  450 FORMAT('0','Enter 1 for display of leakage rates at nodes'/1X,
     *'or 2 to continue program  '\)
      READ(*,*)MS2
      IF(MS2.NE.1.AND.MS2.NE.2)GOTO 440
      IF(MS2.EQ.1)THEN
  458 WRITE(*,460)
  460 FORMAT('0','Enter integer I-coordinate of unit'/1X,
     *'node  '\)
      READ(*,*)I
      WRITE(*,470)
  470 FORMAT(1X,'Enter integer J-coordinate of unit'/1X,
     *'node  '\)
      READ(*,*)J
      WRITE(*,472)
  472 FORMAT(1X,'Enter integer number of unit containing node  '\)
      READ(*,*)K
      IF(K.EQ.1.AND.K-NL.EQ.0)THEN
      ALURAT=0.0
      IF(MUS.EQ.2)ALDRAT=0.0
      IF(MUS.EQ.1)ALDRAT=R(I,J,K)*(H(I,J,K)-HSB)
      GOTO 473
      ENDIF
      IF(K.EQ.1.AND.K-NL.NE.0)THEN
      ALURAT=R(I,J,K)*(H(I,J,K+1)-H(I,J,K))
      IF(MUS.EQ.2)ALDRAT=0.0
      IF(MUS.EQ.1)ALDRAT=R(I,J,K)*(H(I,J,K)-HSB)
      GOTO 473
      ENDIF
      IF(K-NL.EQ.0)THEN
      ALURAT=0.0
      ALDRAT=R(I,J,NL)*(H(I,J,NL-1)-H(I,J,NL))
      GOTO 473
      ENDIF
      ALURAT=R(I,J,K)*(H(I,J,K+1)-H(I,J,K))
```

```
      ALDRAT=R(I,J,K)*(H(I,J,K)-H(I,J,K-1))
  473 ALURAT=ALURAT/1440
      ALDRAT=ALDRAT/1440
      WRITE(*,474)ALURAT
  474 FORMAT(1X,'Leakage rate from below or flow between'/1X,
     *'aquifer layers in gpm=',F9.2)
      WRITE(*,476)ALDRAT
  476 FORMAT(1X,'Leakage rate from above or flow between'/1X,
     *'aquifer layers in gpm=',F9.2)
      WRITE(5,480)K,I,J,TIME(ISTEP),ALURAT
  480 FORMAT(1X,'Leakage or flow rate from below',I2,/1X,
     *'with I-coordinate=',I3,' and J-coordinate=',I3,' at'/1X,
     *'time=',F10.3,' days is',F9.2,' gpm')
      WRITE(5,482)K,I,J,TIME(ISTEP),ALDRAT
  482 FORMAT(1X,'Leakage or flow rate from above',I2,/1X,
     *'with I-coordinate=',I3,' and J-coordinate=',I3,' at'/1X,
     *'time=',F10.3,' days is',F9.2,' gpm')
  490 WRITE(*,500)
  500 FORMAT('0','Enter 1 for another leakage or flow'/1X,
     *'rate or 2 to continue program  '\)
      READ(*,*)MII
      IF(MII.NE.1.AND.MII.NE.2)GOTO 490
      IF(MII.EQ.1)GOTO 458
      ENDIF
  502 WRITE(*,503)
  503 FORMAT('0','Enter 1 for display of aquifer storage'/1X,
     *'flow at specified unit nodes or 2 to continue'/1X,
     *'program  '\)
      READ(*,*)MA1
      IF(MA1.NE.1.AND.MA1.NE.2)GOTO 502
      IF(MA1.EQ.2)GOTO 520
  504 WRITE(*,506)
  506 FORMAT('0','Enter integer I-coordinate of unit node  '\)
      READ(*,*)I
      WRITE(*,508)
  508 FORMAT(1X,'Enter integer J-coordinate of unit node  '\)
      READ(*,*)J
      WRITE(*,510)
  510 FORMAT(1X,'Enter integer unit number containing'/1X,
     *'specified node  '\)
      READ(*,*)K
      S=SF1(I,J,K)
      IF(MDY.EQ.1)S=SF2(I,J,K)
      IF(H(I,J,K).LT.CH(I,J,K))S=SF2(I,J,K)
      STFLO=(S*ABS(HO(I,J,K)-H(I,J,K))/DELTA(ISTEP))/1440
      IF(MDY.EQ.1.AND.K.EQ.1)STFLO=STFLO+QDY(I,J,K)/1440
      WRITE(*,512)STFLO
  512 FORMAT(1X,'Flow from aquifer storage at node in gpm=',F9.2)
      WRITE(5,514)K,I,J,TIME(ISTEP),STFLO
  514 FORMAT(1X,'Flow in unit',I2,' from aquifer'/1X,
     *'storage at node with I-coordinate=',I3,' and'/1X,
     *'J-coordinate=',I3,' at time=',F10.3,' days is',/1X,
```

```
    *F9.2,' gpm')
516 WRITE(*,518)
518 FORMAT('0','Enter 1 for another nodal aquifer'/1X,
    *'storage flow display or 2 to continue program  '\)
     READ(*,*)MA2
     IF(MA2.NE.1.AND.MA2.NE.2)GOTO 516
     IF(MA2.EQ.1)GOTO 504
520 IF(MAS.EQ.1)THEN
522 WRITE(*,524)
524 FORMAT('0','Enter 1 for display of aquitard'/1X,
    *'storage flow at specified nodes or 2 to continue '/1X,
    *'program  '\)
     READ(*,*)MA3
     IF(MA3.NE.1.AND.MA3.NE.2)GOTO 522
     IF(MA3.EQ.2)GOTO 542
526 WRITE(*,528)
528 FORMAT('0','Enter integer I-coordinate of unit node  '\)
     READ(*,*)I
     WRITE(*,530)
530 FORMAT(1X,'Enter integer J-coordinate of unit node  '\)
     READ(*,*)J
     WRITE(*,532)
532 FORMAT(1X,'Enter integer unit number of specified node  '\)
     READ(*,*)K
     STRATE=(RS(I,J,K)*ABS(HO(I,J,K)-H(I,J,K)))/1440
     WRITE(*,534)STRATE
534 FORMAT(1X,'Flow from aquitard storage at node in gpm=',F9.2)
     WRITE(5,536)K,I,J,TIME(ISTEP),STRATE
536 FORMAT(1X,'Flow in unit',I2,' from aquitard storage'/1X,
    *'at node with I-coordinate=',I3,' and J-coordinate=',/1X,
    *I3,' at time=',F10.3,' days is',F9.2,' gpm')
538 WRITE(*,540)
540 FORMAT('0','Enter 1 for another nodal aquitard'/1X,
    *'storage flow display or 2 to continue program  '\)
     READ(*,*)MA4
     IF(MA4.NE.1.AND.MA4.NE.2)GOTO 538
     IF(MA4.EQ.1)GOTO 526
542 ENDIF
     IF(MAP1.EQ.1)THEN
     HMA=A7/A8-QMA/(2*3.1416*A8)
     WRITE(*,546)HMA
546 FORMAT('0','Composite water level or drawdown'/1X,
    *'in multi-aquifer production well in ft=',F9.2)
     DO 550 N=KUM,KLM
     WRITE(*,548)N,Q(IMP,JMP,N)/1440
548 FORMAT(1X,'Discharge from aquifer',I2,' into'/1X,
    *'multi-aquifer production well in gpm=',F9.2)
550 CONTINUE
     ENDIF
     IF(MS1.EQ.1)THEN
640 WRITE(*,650)
650 FORMAT('0','Enter 1 for display of induced infiltration'/1X,
```

```
     *'rates at streambed nodes or 2 to continue program  '\)
      READ(*,*)MS2
      IF(MS2.NE.1.AND.MS2.NE.2)GOTO 640
      IF(MS2.EQ.1)THEN
  658 WRITE(*,660)
  660 FORMAT('0','Enter integer I-coordinate of streambed'/1X,
     *'node  '\)
      READ(*,*)I
      WRITE(*,670)
  670 FORMAT(1X,'Enter integer J-coordinate of streambed'/1X,
     *'node  '\)
      READ(*,*)J
      WRITE(*,680)QST(I,J,1)
  680 FORMAT(1X,'Induced infiltration rate at node in'/1X,
     *'gpd=',F14.2)
      SRATE=QST(I,J,1)/1440
      WRITE(5,682)I,J,TIME(ISTEP),SRATE
  682 FORMAT(1X,'Induced infiltration rate at node with'/1X,
     *'I-coordinate=',I3,' and J-coordinate=',I3,' at'/1X,
     *'time=',F10.3,' days is',F9.2,' gpm')
  690 WRITE(*,700)
  700 FORMAT('0','Enter 1 for another induced infiltration'/1X,
     *'rate or 2 to continue program  '\)
      READ(*,*)MII
      IF(MII.NE.1.AND.MII.NE.2)GOTO 690
      IF(MII.EQ.1)GOTO 658
      ENDIF
      ENDIF
      IF(MD1.EQ.1)THEN
  710 WRITE(*,720)
  720 FORMAT('0','Enter 1 for display of drainage rates at'/1X,
     *'drain nodes or 2 to continue program  '\)
      READ(*,*)MD2
      IF(MD2.NE.1.AND.MD2.NE.2)GOTO 710
      IF(MD2.EQ.1)THEN
  728 WRITE(*,730)
  730 FORMAT('0','Enter integer I-coordinate of drain node  '\)
      READ(*,*)I
      WRITE(*,740)
  740 FORMAT(1X,'Enter integer J-coordinate of drain node  '\)
      READ(*,*)J
      DRTO=QDRAIN(I,J,1)/1440
      WRITE(*,760)DRTO
  760 FORMAT(1X,'Drainage rate at drain node in gpm=',F9.2)
      WRITE(5,762)I,J,TIME(ISTEP),DRTO
  762 FORMAT(1X,'Discharge rate from drain at node with'/1X,
     *'I-coordinate=',I3,' and J-coordinate=',I3,' at'/1X,
     *'time',F10.3,' days is',F9.2,' gpm')
  770 WRITE(*,780)
  780 FORMAT('0','Enter 1 for another drainage rate or 2'/1X,
     *'to continue program  '\)
      READ(*,*)MD3
```

```
      IF(MD3.NE.1.AND.MD3.NE.2)GOTO 770
      IF(MD3.EQ.1)GOTO 728
      ENDIF
      ENDIF
790   CLOSE (5)
      END
```

```
C
      SUBROUTINE RIVER(R,RH,RD,MSB,MS1,MSN)
C
C     THIS SUBROUTINE ENTERS AND DISPLAYS RIVER
C     SOURCE GEOMETRY AND STRENGTH DATA BASE
C
C     DECLARATIONS
C
      DIMENSION R(20,20,4),RH(20,20,1),RD(20,20,1),MSB(20,20,1)
C
    5 WRITE(*,7)
    7 FORMAT('0','Enter 1 to simulate induced streambed'/1X,
     *'infiltration in aquifer unit or layer number 1 or'/1X,
     *'2 to continue program  '\)
      READ(*,*)MS1
      IF(MS1.NE.1.AND.MS1.NE.2)GOTO 5
      IF(MS1.EQ.1)THEN
      OPEN(5,FILE='LPT1:')
    8 WRITE(*,10)
   10 FORMAT('0','Enter integer number of stream nodes in'/1X,
     *'uppermost aquifer unit or layer  '\)
      READ(*,*)MSN
      DO 200 N=1,MSN
      WRITE(*,20)N
   20 FORMAT('0','Stream node number=',I3)
      WRITE(5,30)N
   30 FORMAT('0','Stream node number=',I3)
      WRITE(*,40)
   40 FORMAT(1X,'Enter integer I-coordinate of stream node  '\)
      READ(*,*)I
      WRITE(5,50)I
   50 FORMAT(1X,'I-coordinate of stream node=',I3)
      WRITE(*,60)
   60 FORMAT(1X,'Enter integer J-coordinate of stream node  '\)
      READ(*,*)J
      WRITE(5,70)J
   70 FORMAT(1X,'J-coordinate of stream node=',I3)
      WRITE(*,80)
   80 FORMAT(1X,'Enter real number streambed width'/1X,
     *'simulated at node in ft  '\)
      READ(*,*)SBW
      WRITE(5,90)SBW
   90 FORMAT(1X,'Streambed width simulated at node in ft=',F7.2)
      WRITE(*,100)
  100 FORMAT(1X,'Enter real number streambed length'/1X,
     *'simulated at node in ft  '\)
```

```
      READ(*,*)SBL
      WRITE(5,110)SBL
110 FORMAT(1X,'Streambed length simulated at node in ft=',F9.2)
112 WRITE(*,114)
114 FORMAT('0','Enter 1 to specify induced streambed'/1X,
    *'infiltration rate or 2 to specify streambed vertical'/1X,
    *'hydraulic conductivity and thickness  '\)
      READ(*,*)MIS
      IF(MIS.NE.1.AND.MIS.NE.2)GOTO 112
      IF(MIS.EQ.1)THEN
      WRITE(*,115)
115 FORMAT('0','Enter real number induced streambed'/1X,
    *'infiltration rate in gpd per acre of streambed'/1X,
    *'per foot of head loss  '\)
      READ(*,*)SIR
      WRITE(5,116)SIR
116 FORMAT(1X,'Induced streambed infiltration rate'/1X,
    *'in gpd per acre per ft=',F9.2)
      R(I,J,1)=(SIR/43560)*SBW*SBL
      GOTO 154
      ENDIF
      WRITE(*,120)
120 FORMAT(1X,'Enter real number streambed thickness'/1X,
    *'simulated at node in ft, usually ranges from .1 to 5 ft  '\)
      READ(*,*)SBT
      WRITE(5,130)SBT
130 FORMAT(1X,'Streambed thickness simulated at node in ft=',F5.2)
      WRITE(*,140)
140 FORMAT(1X,'Enter real number hydraulic conductivity'/1X,
    *'of streambed simulated at node in gpd/sq ft, usually'/1X,
    *'ranges from 1E-2 to 1000 gpd/sq ft  '\)
      READ(*,*)SBH
      WRITE(5,150)SBH
150 FORMAT(1X,'Hydraulic conductivity of streambed'/1X,
    *'simulated at node in gpd/sq ft=',F8.3)
      R(I,J,1)=(SBH/SBT)*SBW*SBL
154 WRITE(*,160)
160 FORMAT(1X,'Enter real number stream surface'/1X,
    *'elevation in ft  '\)
      READ(*,*)RH(I,J,1)
      WRITE(5,170)RH(I,J,1)
170 FORMAT(1X,'Stream surface elevation in ft=',F9.2)
      WRITE(*,180)
180 FORMAT(1X,'Enter real number streambed base'/1X,
    *'elevation in ft  '\)
      READ(*,*)RD(I,J,1)
      WRITE(5,190)RD(I,J,1)
190 FORMAT(1X,'Streambed base elevation in ft=',F9.2)
      MSB(I,J,1)=1
200 CONTINUE
210 WRITE(*,220)
220 FORMAT('0','Enter 1 to revise stream data base or'/1X,
```

```
   *'2 to continue program  '\)
    READ(*,*)MS2
    IF(MS2.NE.1.AND.MS2.NE.2)GOTO 210
    IF(MS2.EQ.1)GOTO 8
    CLOSE (5)
    ENDIF
230 END
```

```
C
      SUBROUTINE SIMEQ(H,HO,SF1,Q,T1,T2,NC,NR,NL,DELTA,
     *ISTEP,R,DL,RS,MAS,DELX,DELY,MDY,SF2,MAP1,IMP,JMP,
     *A7,A8,QMA,KUM,KLM,ERROR,S1,S2,ALPHA,TIME,QDY,HD,
     *E,ITER,WBE,NIC,QP,QI,QR,BOT,CH,MFW,CELE,MSB,RD,RH,
     *MDR,P1,P2,MF1,MDM1,MS1,MD1,ME1,QF,QRD,QST,QDRAIN,QE,
     *MUS,HSB,G,B,MQE,MF2,MAU)
C
C     THIS SUBROUTINE SOLVES FINITE-DIFFERENCE SIMULTANEOUS
C     EQUATIONS USING THE ITERATIVE ALTERNATING DIRECTION
C     IMPLICIT METHOD AND ENTERS SOME PARTS OF DATA BASE
C
C     METACOMMANDS
C
$LARGE
C
C     DECLARATIONS
C
      DIMENSION H(20,20,4),HO(20,20,4),DL(20,20,4),SF1(20,20,4),
     *Q(20,20,4),R(20,20,4),T1(20,20,4),T2(20,20,4),B(21,4),
     *G(21,4),DELTA(50),RS(20,20,4),SF2(20,20,4),DELX(21),
     *DELY(21),QDY(20,20,1),A2(4),S1(20,20,4),S2(20,20,4),
     *ALPHA(20,20,1),TIME(50),HD(20,20,20),QP(20,20,4),QI(20,20,4),
     *QR(20,20,1),BOT(20,20,4),CH(20,20,4),MFW(20,20,4),MSB(20,20,1),
     *CELE(20,20,4),RD(20,20,1),RH(20,20,1),P1(20,20,4),P2(20,20,4),
     *MDR(20,20,1),QF(20,20,4),QRD(20,20,4),QST(20,20,4),
     *QDRAIN(20,20,1),QE(20,20,1),MQE(20,20,1),RMA(4),A1(4)
      DO 8 K=1,NL
      DO 7 J=1,NR
      DO 6 I=1,NC
      Q(I,J,K)=QP(I,J,K)-QI(I,J,K)
      IF(K.EQ.1)Q(I,J,K)=Q(I,J,K)-QR(I,J,K)+QE(I,J,K)
    6 CONTINUE
    7 CONTINUE
    8 CONTINUE
C
C     PREDICT HEADS FOR NEXT TIME INCREMENT
C
      DO 12 K=1,NL
      DO 12 I=1,NC
      DO 12 J=1,NR
      IF(SF1(I,J,K).EQ.1E21.OR.SF2(I,J,K).EQ.1E21)GOTO 11
      D=H(I,J,K)-HO(I,J,K)
      HO(I,J,K)=H(I,J,K)
      F=1.0
      IF(DL(I,J,K).EQ.0.0)GOTO 10
```

```
      IF(ISTEP.GT.2)F=D/DL(I,J,K)
      IF(F.GT.5)F=5.0
      IF(F.LT.0.0)F=0.0
   10 DL(I,J,K)=D
      H(I,J,K)=H(I,J,K)+D*F
   11 IF(H(I,J,K).LE.BOT(I,J,K))H(I,J,K)=BOT(I,J,K)+0.01
   12 CONTINUE
C
C     CALCULATE EFFECTIVE STORATIVITY AND RECHARGE
C     UNDER DELAYED GRAVITY YIELD CONDITIONS
C
      IF(MDY.EQ.1)THEN
      SUM1=0.0
      DO 16 J=1,NR
      DO 15 I=1,NC
      SF2(I,J,1)=S1(I,J,1)+S2(I,J,1)*(1-EXP(-ALPHA(I,J,1)*
     *TIME(ISTEP)))
      SF2(I,J,1)=SF2(I,J,1)*7.48*DELX(I)*DELY(J)
      IF(ISTEP.EQ.1)THEN
      QDY(I,J,1)=0.0
      GOTO 15
      ENDIF
      DO 14 L=1,ISTEP-1
      F2=EXP(-ALPHA(I,J,1)*(TIME(ISTEP)-TIME(L)))
      SUM1=SUM1+HD(I,J,L)*F2
   14 CONTINUE
      QDY(I,J,1)=ALPHA(I,J,1)*S2(I,J,1)*SUM1*7.48*DELX(I)*
     *DELY(J)
      SUM1=0.0
   15 CONTINUE
   16 CONTINUE
   17 ENDIF
      IF(MDM1.EQ.1.AND.ISTEP.EQ.1)THEN
   18 WRITE(*,19)
   19 FORMAT('0','Enter 1 if there are mine nodes in the upper'/1X,
     *'most aquifer unit or layer and water table conditions'/1X,
     *'occur in the uppermost aquifer or layer. Enter 2 if there'/1X,
     *'are no mine nodes in the uppermost aquifer unit or layer'/1X,
     *'or there are mine nodes in the uppermost aquifer unit or'/1X,
     *'layer but leaky or nonleaky artesian conditions occur in'/1X,
     *'the uppermost aquifer unit or layer   '\)
      READ(*,*)MTT
      IF(MTT.NE.1.AND.MTT.NE.2)GOTO 18
      ENDIF
C
C     ENTER MULTI-UNIT OR LAYER PRODUCTION WELL DATA BASE
C
      IF(ISTEP.GT.1)GOTO 22
   20 WRITE(*,21)
   21 FORMAT('0','Enter 1 to simulate a multi-unit or layer'/1X,
     *'production well or 2 to continue program. The model'/1X,
     *'assumes a single multi-unit or layer production'/1X,
```

```
     *'well open to selected adjacent units or layers in the'/1X,
     *'system. The radius of the production well is the same in'/1X,
     *'all units or layers.  '\)
      READ(*,*)MAP1
      IF(MAP1.NE.1.AND.MAP1.NE.2)GOTO 20
   22 IF(MAP1.EQ.1.AND.ISTEP.LT.2)THEN
      WRITE(*,23)
   23 FORMAT('0','Enter integer I-coordinate of production'/1X,
     *'well  '\)
      READ(*,*)IMP
      WRITE(*,24)
   24 FORMAT(1X,'Enter integer J-coordinate of production'/1X,
     *'well  '\)
      READ(*,*)JMP
      WRITE(*,25)
   25 FORMAT(1X,'Enter integer uppermost unit open to'/1X,
     *'production well  '\)
      READ(*,*)KUM
      WRITE(*,26)
   26 FORMAT(1X,'Enter integer lowermost unit open to'/1X,
     *'production well  '\)
      READ(*,*)KLM
      WRITE(*,27)
   27 FORMAT(1X,'Enter real number production well radius'/1X,
     *'in ft  '\)
      READ(*,*)PMA
      DO 28 N=KUM,KLM
      RMA(N)=0.28*((T1(IMP,JMP,N)/T2(IMP,JMP,N))**.5*DELX(IMP)**2+
     *(T2(IMP,JMP,N)/T1(IMP,JMP,N))**.5*DELY(JMP)**2)**.5/
     *((T1(IMP,JMP,N)/T2(IMP,JMP,N))**.25+
     *(T2(IMP,JMP,N)/T1(IMP,JMP,N))**.25)
      A1(N)=ALOG(RMA(N)/PMA)
   28 CONTINUE
      ENDIF
      IF(MAP1.EQ.1)THEN
      WRITE(*,29)
   29 FORMAT(1X,'Enter discharge rate from multi-'/1X,
     *'unit or layer production well in gpm. The discharge'/1X,
     *'rate may be zero in which case there will be flow'/1X,
     *'with head differences from one unit or layer to'/1X,
     *'another. The discharge rate may be negative to'/1X,
     *'simulate a recharging well  '\)
      READ(*,*)QMA
      QMA=QMA*1440
      ENDIF
      WRITE(*,30)
   30 FORMAT('0','COMPUTATIONS ARE IN PROGRESS')
C
C     REFINE ESTIMATES OF HEADS
C
C     INITIALIZE ERROR AND ITERATION COUNTER
C
```

```
      ITER=0
   31 E=0.0
      ITER=ITER+1
C
C     CALCULATE VALUES OF Q(I,J,K) IN EACH UNIT OR LAYER
C     ASSOCIATED WITH MULTI-UNIT OR LAYER PRODUCTION WELL
C
      IF(MAP1.EQ.1)THEN
      A7=0.0
      A8=0.0
      DO 32 N=KUM,KLM
      A2(N)=(T1(IMP,JMP,N)*T2(IMP,JMP,N))**.5
      A7=A7+(A2(N)*H(IMP,JMP,N))/A1(N)
      A8=A8+A2(N)/A1(N)
   32 CONTINUE
      DO 33 N=KUM,KLM
      A5=2*3.1416*A2(N)*H(IMP,JMP,N)/A1(N)
      A6=2*3.1416*A2(N)/A1(N)
      Q(IMP,JMP,N)=A5-A6*(A7/A8)+A2(N)*QMA/(A1(N)*A8)
   33 CONTINUE
      ENDIF
C     UNIT AND COLUMN CALCULATIONS
C
      DO 160 K=1,NL
      DO 160 I=1,NC
      DO 140 J=1,NR
C
C     CALCULATE PEACEMAN-RACHFORD B AND G ARRAYS
C
      IF(K.NE.1)HR=H(I,J,K-1)
      S=SF1(I,J,K)
      IF(H(I,J,K).LT.BOT(I,J,K))H(I,J,K)=BOT(I,J,K)
      IF(H(I,J,K).LT.CH(I,J,K))S=SF2(I,J,K)
      IF(MDY.EQ.1)S=SF2(I,J,K)
C
C     SIMULATE SINK/SOURCE TERMS
C
      IF(K.EQ.1)HR=HSB
      IF(MAS.EQ.1)THEN
      IF(K.EQ.1.AND.MUS.EQ.2)GOTO 37
      IF(MAU.EQ.1.OR.MAU.EQ.2.OR.MAU.EQ.3)GOTO 37
      IF(MFW(I,J,K).EQ.1)THEN
      IF(MF2.EQ.1)R(I,J,K)=0.0
      RE=CELE(I,J,K)*R(I,J,K)
      DD=HO(I,J,K)*S/DELTA(ISTEP)-Q(I,J,K)+RE+HO(I,J,K)*RS(I,J,K)
      BB=S/DELTA(ISTEP)+R(I,J,K)+RS(I,J,K)
      GOTO 36
      ENDIF
      IF(H(I,J,K).LT.CH(I,J,K))THEN
      DD=HO(I,J,K)*S/DELTA(ISTEP)-Q(I,J,K)+
     *R(I,J,K)*(HR-CH(I,J,K))
      IF((HO(I,J,K)-CH(I,J,K))*(H(I,J,K)-CH(I,J,K)).LT.0.0)
```

```
  *DD=DD+(HO(I,J,K)-CH(I,J,K))*(SF1(I,J,K)-SF2(I,J,K))
  */DELTA(ISTEP)
   BB=S/DELTA(ISTEP)+R(I,J,K)
   GOTO 48
   ENDIF
34 DD=HO(I,J,K)*S/DELTA(ISTEP)-Q(I,J,K)+
  *R(I,J,K)*HR+HO(I,J,K)*RS(I,J,K)
   BB=S/DELTA(ISTEP)+R(I,J,K)+RS(I,J,K)
36 GOTO 48
   ENDIF
37 IF(MDY.EQ.1.AND.K.EQ.1)THEN
   IF(MSB(I,J,1).EQ.1.OR.MQE(I,J,1).EQ.1)THEN
   IF(H(I,J,K).LT.RD(I,J,1))GOTO 38
   RE=RH(I,J,1)*R(I,J,K)
   RB=1.0
   GOTO 39
38 RE=(RH(I,J,1)-RD(I,J,1))*R(I,J,K)
   RB=0.0
39 DD=HO(I,J,K)*S/DELTA(ISTEP)-Q(I,J,K)+QDY(I,J,K)+RE
   BB=S/DELTA(ISTEP)+R(I,J,K)*RB
   ENDIF
   IF(MDR(I,J,1).EQ.1)THEN
   IF(H(I,J,K).LT.RH(I,J,K))R(I,J,K)=0.0
   RE=RH(I,J,K)*R(I,J,K)
   DD=HO(I,J,K)*S/DELTA(ISTEP)-Q(I,J,K)+QDY(I,J,K)+RE
   BB=S/DELTA(ISTEP)+R(I,J,K)
   ENDIF
   IF(MSB(I,J,1).EQ.1.OR.MDR(I,J,1).EQ.1.OR.MQE(I,J,1).EQ.1)GOTO 41
   DD=HO(I,J,K)*S/DELTA(ISTEP)-Q(I,J,K)+QDY(I,J,K)+
  *R(I,J,K)*HR
   BB=S/DELTA(ISTEP)+R(I,J,K)
40 GOTO 48
   ENDIF
41 IF(MFW(I,J,K).EQ.1)THEN
   IF(MF2.EQ.1)R(I,J,K)=0.0
   RE=CELE(I,J,K)*R(I,J,K)
   DD=HO(I,J,K)*S/DELTA(ISTEP)-Q(I,J,K)+RE
   BB=S/DELTA(ISTEP)+R(I,J,K)
   ENDIF
   IF(K.EQ.1)THEN
   IF(MSB(I,J,1).EQ.1.OR.MQE(I,J,1).EQ.1)THEN
   IF(H(I,J,K).LT.RD(I,J,K))GOTO 42
   RE=RH(I,J,K)*R(I,J,K)
   RB=1.0
   GOTO 43
42 RE=(RH(I,J,K)-RD(I,J,K))*R(I,J,K)
   RB=0.0
43 DD=HO(I,J,K)*S/DELTA(ISTEP)-Q(I,J,K)+RE
   BB=S/DELTA(ISTEP)+R(I,J,K)*RB
   ENDIF
   IF(MDR(I,J,1).EQ.1)THEN
   IF(H(I,J,K).LT.RH(I,J,K))R(I,J,K)=0.0
```

```
      RE=RH(I,J,K)*R(I,J,K)
      DD=HO(I,J,K)*S/DELTA(ISTEP)-Q(I,J,K)+RE
      BB=S/DELTA(ISTEP)+R(I,J,K)
      ENDIF
      ENDIF
      IF(MFW(I,J,K).EQ.1)GOTO 48
      IF(K.EQ.1.AND.MSB(I,J,1).EQ.1)GOTO 48
      IF(K.EQ.1.AND.MDR(I,J,1).EQ.1)GOTO 48
      IF(K.EQ.1.AND.MQE(I,J,1).EQ.1)GOTO 48
   47 DD=HO(I,J,K)*S/DELTA(ISTEP)-Q(I,J,K)+R(I,J,K)*HR
      BB=S/DELTA(ISTEP)+R(I,J,K)
      IF(K.EQ.1.AND.MUS.EQ.2)GOTO 48
      IF(H(I,J,K).LT.CH(I,J,K))THEN
      DD=HO(I,J,K)*S/DELTA(ISTEP)-Q(I,J,K)+
     *R(I,J,K)*(HR-CH(I,J,K))
      IF((HO(I,J,K)-CH(I,J,K))*(H(I,J,K)-CH(I,J,K)).LT.0.0)
     *DD=DD+(HO(I,J,K)-CH(I,J,K))*(SF1(I,J,K)-SF2(I,J,K))
     */DELTA(ISTEP)
      BB=S/DELTA(ISTEP)+R(I,J,K)
      ENDIF
   48 AA=0.0
      CC=0.0
      IF(K-NL.EQ.0.0)GOTO 50
      IF(MFW(I,J,K+1).EQ.1)R(I,J,K+1)=0.0
      IF(H(I,J,K+1).LE.CH(I,J,K+1))THEN
      DD=DD+R(I,J,K+1)*CH(I,J,K+1)
      IF((HO(I,J,K)-CH(I,J,K))*(H(I,J,K)-CH(I,J,K)).LT.0.0)
     *DD=DD+(HO(I,J,K)-CH(I,J,K))*(SF1(I,J,K)-SF2(I,J,K))
     */DELTA(ISTEP)
      GOTO 49
      ENDIF
      BB=BB+R(I,J,K+1)
      DD=DD+R(I,J,K+1)*H(I,J,K+1)
   49 IF(MFW(I,J,K+1).NE.1.AND.MAS.EQ.1)THEN
      BB=BB+RS(I,J,K+1)
      DD=DD+HO(I,J,K+1)*RS(I,J,K+1)
      ENDIF
   50 IF(J-1)60,70,60
   60 AA=-T1(I,J-1,K)
      BB=BB+T1(I,J-1,K)
   70 IF(J-NR)80,90,80
   80 IF(H(I,J,K).LE.CH(I,J,K))THEN
      IF(P1(I,J,K).EQ.0.0.OR.MTT.EQ.2)GOTO 82
      IF(SF1(I,J,K).EQ.1E21.OR.SF2(I,J,K).EQ.1E21.OR.
     *SF1(I,J+1,K).EQ.1E21.OR.SF2(I,J+1,K).EQ.1E21)THEN
      T1(I,J,K)=P1(I,J,K)*0.5*((AMIN1(H(I,J,K),CH(I,J,K))
     *-BOT(I,J,K))+(AMIN1(H(I,J+1,K),CH(I,J+1,K))-BOT(I,J+1,K)))
      GOTO 82
      ENDIF
      T1(I,J,K)=P1(I,J,K)*SQRT((AMIN1(H(I,J,K),CH(I,J,K))
     *-BOT(I,J,K))*(AMIN1(H(I,J+1,K),CH(I,J+1,K))-BOT(I,J+1,K)))
      ENDIF
```

```
 82 CC=-T1(I,J,K)
    BB=BB+T1(I,J,K)
 90 IF(I-1)100,110,100
100 BB=BB+T2(I-1,J,K)
    DD=DD+H(I-1,J,K)*T2(I-1,J,K)
110 IF(I-NC)120,130,120
120 IF(H(I,J,K).LE.CH(I,J,K))THEN
    IF(P2(I,J,K).EQ.0.0.OR.MTT.EQ.2)GOTO 122
    IF(SF1(I,J,K).EQ.1E21.OR.SF2(I,J,K).EQ.1E21.OR.
   *SF1(I+1,J,K).EQ.1E21.OR.SF2(I+1,J,K).EQ.1E21)THEN
    T2(I,J,K)=P2(I,J,K)*0.5*((AMIN1(H(I,J,K),CH(I,J,K))-
   *BOT(I,J,K))+(AMIN1(H(I+1,J,K),CH(I+1,J,K))-BOT(I+1,J,K)))
    GOTO 122
    ENDIF
    T2(I,J,K)=P2(I,J,K)*SQRT((AMIN1(H(I,J,K),CH(I,J,K))-
   *BOT(I,J,K))*(AMIN1(H(I+1,J,K),CH(I+1,J,K))-BOT(I+1,J,K)))
    ENDIF
122 BB=BB+T2(I,J,K)
    DD=DD+H(I+1,J,K)*T2(I,J,K)
130 IF(J.GT.1)W=BB-AA*B(J-1,K)
    IF(J.EQ.1)W=BB
    B(J,K)=CC/W
    IF(J.EQ.1)G(J,K)=DD/W
    IF(J.GT.1)G(J,K)=(DD-AA*G(J-1,K))/W
140 CONTINUE
C
C     RE-ESTIMATE HEADS
C
    E=E+ABS(H(I,NR,K)-G(NR,K))
    H(I,NR,K)=G(NR,K)
    N=NR-1
150 HA=G(N,K)-B(N,K)*H(I,N+1,K)
    E=E+ABS(HA-H(I,N,K))
    H(I,N,K)=HA
    N=N-1
    IF(N.GT.0)GOTO 150
    DO 155 N=1,NR
    IF(H(I,N,K).GT.BOT(I,N,K)) GOTO 155
    E=E+BOT(I,N,K)+0.01-H(I,N,K)
    H(I,N,K)=BOT(I,N,K)+0.01
155 CONTINUE
160 CONTINUE
C
C     UNIT AND ROW CALCULATIONS
C
    DO 290 K=1,NL
    DO 290 J=1,NR
    DO 270 I=1,NC
    IF(K.NE.1)HR=H(I,J,K-1)
    S=SF1(I,J,K)
    IF(H(I,J,K).LT.BOT(I,J,K))H(I,J,K)=BOT(I,J,K)
    IF(H(I,J,K).LT.CH(I,J,K))S=SF2(I,J,K)
```

```
      IF(MDY.EQ.1)S=SF2(I,J,K)
C
C     SIMULATE SINK/SOURCE TERMS
C
      IF(K.EQ.1)HR=HSB
      IF(MAS.EQ.1)THEN
      IF(K.EQ.1.AND.MUS.EQ.2)GOTO 163
      IF(MAU.EQ.1.OR.MAU.EQ.2.OR.MAU.EQ.3)GOTO 163
      IF(MFW(I,J,K).EQ.1)THEN
      IF(MF2.EQ.1)R(I,J,K)=0.0
      RE=CELE(I,J,K)*R(I,J,K)
      DD=HO(I,J,K)*S/DELTA(ISTEP)-Q(I,J,K)+RE+HO(I,J,K)*RS(I,J,K)
      BB=S/DELTA(ISTEP)+R(I,J,K)+RS(I,J,K)
      GOTO 162
      ENDIF
      IF(H(I,J,K).LT.CH(I,J,K))THEN
      DD=HO(I,J,K)*S/DELTA(ISTEP)-Q(I,J,K)+
     *R(I,J,K)*(HR-CH(I,J,K))
      IF((HO(I,J,K)-CH(I,J,K))*(H(I,J,K)-CH(I,J,K)).LT.0.0)
     *DD=DD+(HO(I,J,K)-CH(I,J,K))*(SF1(I,J,K)-SF2(I,J,K))
     */DELTA(ISTEP)
      BB=S/DELTA(ISTEP)+R(I,J,K)
      GOTO 174
      ENDIF
  161 DD=HO(I,J,K)*S/DELTA(ISTEP)-Q(I,J,K)+
     *R(I,J,K)*HR+HO(I,J,K)*RS(I,J,K)
      BB=S/DELTA(ISTEP)+R(I,J,K)+RS(I,J,K)
  162 GOTO 174
      ENDIF
  163 IF(MDY.EQ.1.AND.K.EQ.1)THEN
      IF(MSB(I,J,1).EQ.1.OR.MQE(I,J,1).EQ.1)THEN
      IF(H(I,J,K).LT.RD(I,J,1))GOTO 164
      RE=RH(I,J,1)*R(I,J,K)
      RB=1.0
      GOTO 166
  164 RE=(RH(I,J,1)-RD(I,J,1))*R(I,J,K)
      RB=0.0
  166 DD=HO(I,J,K)*S/DELTA(ISTEP)-Q(I,J,K)+QDY(I,J,K)+RE
      BB=S/DELTA(ISTEP)+R(I,J,K)*RB
      ENDIF
      IF(MDR(I,J,1).EQ.1)THEN
      IF(H(I,J,K).LT.RH(I,J,K))R(I,J,K)=0.0
      RE=RH(I,J,K)*R(I,J,K)
      DD=HO(I,J,K)*S/DELTA(ISTEP)-Q(I,J,K)+QDY(I,J,K)+RE
      BB=S/DELTA(ISTEP)+R(I,J,K)
      ENDIF
      IF(MSB(I,J,1).EQ.1.OR.MDR(I,J,1).EQ.1.OR.MQE(I,J,1).EQ.1)
     *GOTO 169
      DD=HO(I,J,K)*S/DELTA(ISTEP)-Q(I,J,K)+QDY(I,J,K)+
     *R(I,J,K)*HR
      BB=S/DELTA(ISTEP)+R(I,J,K)
  168 GOTO 174
```

```
      ENDIF
 169  IF(MFW(I,J,K).EQ.1)THEN
      IF(MF2.EQ.1)R(I,J,K)=0.0
      RE=CELE(I,J,K)*R(I,J,K)
      DD=HO(I,J,K)*S/DELTA(ISTEP)-Q(I,J,K)+RE
      BB=S/DELTA(ISTEP)+R(I,J,K)
      ENDIF
      IF(K.EQ.1)THEN
      IF(MSB(I,J,1).EQ.1.OR.MQE(I,J,1).EQ.1)THEN
      IF(H(I,J,K).LT.RD(I,J,K))GOTO 170
      RE=RH(I,J,K)*R(I,J,K)
      RB=1.0
      GOTO 171
 170  RE=(RH(I,J,K)-RD(I,J,K))*R(I,J,K)
      RB=0.0
 171  DD=HO(I,J,K)*S/DELTA(ISTEP)-Q(I,J,K)+RE
      BB=S/DELTA(ISTEP)+R(I,J,K)*RB
      ENDIF
      IF(MDR(I,J,1).EQ.1)THEN
      IF(H(I,J,K).LT.RH(I,J,K))R(I,J,K)=0.0
      RE=RH(I,J,K)*R(I,J,K)
      DD=HO(I,J,K)*S/DELTA(ISTEP)-Q(I,J,K)+RE
      BB=S/DELTA(ISTEP)+R(I,J,K)
      ENDIF
      ENDIF
      IF(MFW(I,J,K).EQ.1)GOTO 174
      IF(K.EQ.1.AND.MSB(I,J,1).EQ.1)GOTO 174
      IF(K.EQ.1.AND.MDR(I,J,1).EQ.1)GOTO 174
      IF(K.EQ.1.AND.MQE(I,J,1).EQ.1)GOTO 174
 172  DD=HO(I,J,K)*S/DELTA(ISTEP)-Q(I,J,K)+R(I,J,K)*HR
      BB=S/DELTA(ISTEP)+R(I,J,K)
      IF(K.EQ.1.AND.MUS.EQ.2)GOTO 174
      IF(H(I,J,K).LT.CH(I,J,K))THEN
      DD=HO(I,J,K)*S/DELTA(ISTEP)-Q(I,J,K)+
     *R(I,J,K)*(HR-CH(I,J,K))
      IF((HO(I,J,K)-CH(I,J,K))*(H(I,J,K)-CH(I,J,K)).LT.0.0)
     *DD=DD+(HO(I,J,K)-CH(I,J,K))*(SF1(I,J,K)-SF2(I,J,K))
     */DELTA(ISTEP)
      BB=S/DELTA(ISTEP)+R(I,J,K)
      ENDIF
 174  AA=0.0
      CC=0.0
      IF(K-NL.EQ.0.0)GOTO 180
      IF(MFW(I,J,K+1).EQ.1)R(I,J,K+1)=0.0
      IF(H(I,J,K+1).LE.CH(I,J,K+1))THEN
      DD=DD+R(I,J,K+1)*CH(I,J,K+1)
      IF((HO(I,J,K)-CH(I,J,K))*(H(I,J,K)-CH(I,J,K)).LT.0.0)
     *DD=DD+(HO(I,J,K)-CH(I,J,K))*(SF1(I,J,K)-SF2(I,J,K))
     */DELTA(ISTEP)
      GOTO 176
      ENDIF
      BB=BB+R(I,J,K+1)
```

```
      DD=DD+R(I,J,K+1)*H(I,J,K+1)
  176 IF(MFW(I,J,K+1).NE.1.AND.MAS.EQ.1)THEN
      BB=BB+RS(I,J,K+1)
      DD=DD+HO(I,J,K+1)*RS(I,J,K+1)
      ENDIF
  180 IF(J-1)190,200,190
  190 BB=BB+T1(I,J-1,K)
      DD=DD+H(I,J-1,K)*T1(I,J-1,K)
  200 IF(J-NR)210,220,210
  210 IF(H(I,J,K).LE.CH(I,J,K))THEN
      IF(P1(I,J,K).EQ.0.0.OR.MTT.EQ.2)GOTO 212
      IF(SF1(I,J,K).EQ.1E21.OR.SF2(I,J,K).EQ.1E21.OR.
     *SF1(I,J+1,K).EQ.1E21.OR.SF2(I,J+1,K).EQ.1E21)THEN
      T1(I,J,K)=P1(I,J,K)*0.5*((AMIN1(H(I,J,K),CH(I,J,K))-
     *BOT(I,J,K))+(AMIN1(H(I,J+1,K),CH(I,J+1,K))-BOT(I,J+1,K)))
      GOTO 212
      ENDIF
      T1(I,J,K)=P1(I,J,K)*SQRT((AMIN1(H(I,J,K),CH(I,J,K))-
     *BOT(I,J,K))*(AMIN1(H(I,J+1,K),CH(I,J+1,K))-BOT(I,J+1,K)))
      ENDIF
  212 DD=DD+H(I,J+1,K)*T1(I,J,K)
      BB=BB+T1(I,J,K)
  220 IF(I-1)230,240,230
  230 BB=BB+T2(I-1,J,K)
      AA=-T2(I-1,J,K)
  240 IF(I-NC)250,260,250
  250 IF(H(I,J,K).LE.CH(I,J,K))THEN
      IF(P2(I,J,K).EQ.0.0.OR.MTT.EQ.2)GOTO 252
      IF(SF1(I,J,K).EQ.1E21.OR.SF2(I,J,K).EQ.1E21.OR.
     *SF1(I+1,J,K).EQ.1E21.OR.SF2(I+1,J,K).EQ.1E21)THEN
      T2(I,J,K)=P2(I,J,K)*0.5*((AMIN1(H(I,J,K),CH(I,J,K))-
     *BOT(I,J,K))+(AMIN1(H(I+1,J,K),CH(I+1,J,K))-BOT(I+1,J,K)))
      GOTO 252
      ENDIF
      T2(I,J,K)=P2(I,J,K)*SQRT((AMIN1(H(I,J,K),CH(I,J,K))-
     *BOT(I,J,K))*(AMIN1(H(I+1,J,K),CH(I+1,J,K))-BOT(I+1,J,K)))
      ENDIF
  252 BB=BB+T2(I,J,K)
      CC=-T2(I,J,K)
  260 IF(I.GT.1)W=BB-AA*B(I-1,K)
      IF(I.EQ.1)W=BB
      B(I,K)=CC/W
      IF(I.EQ.1)G(I,K)=DD/W
      IF(I.GT.1)G(I,K)=(DD-AA*G(I-1,K))/W
  270 CONTINUE
C
C     RE-ESTIMATE HEADS
C
      E=E+ABS(H(NC,J,K)-G(NC,K))
      H(NC,J,K)=G(NC,K)
      N=NC-1
  280 HA=G(N,K)-B(N,K)*H(N+1,J,K)
```

```
      E=E+ABS(H(N,J,K)-HA)
      H(N,J,K)=HA
      N=N-1
      IF(N.GT.0)GOTO 280
      DO 282 N=1,NC
      IF(H(N,J,K).GT.BOT(N,J,K))GOTO 282
      E=E+BOT(N,J,K)+0.01-H(N,J,K)
      H(N,J,K)=BOT(N,J,K)+0.01
  282 CONTINUE
  290 CONTINUE
C
C     CALCULATE ERROR AND WATER BALANCE OF CONVERGENCE
C     AND DISPLAY PRELIMINARY COMPUTATION RESULTS
C
      CALL WATBL(NC,NR,NL,SF1,H,HO,DELTA,ERROR,E,ITER,
     *WBE,ISTEP,Q,R,NIC,RS,MAS,SF2,QDY,MDY,CH,CELE,QF,QRD,
     *QST,MDR,QDRAIN,QE,RH,RD,T1,T2,QR,QP,MSB,MF1,MDM1,MS1,
     *MD1,ME1,QI,MFW,MUS,HSB,MAP1,IMP,JMP,MAU,BOT)
      IF(ITER.GT.50)GOTO 300
      IF(E.GT.ERROR)GOTO 31
  300 END
```

```
C
      SUBROUTINE WATBL(NC,NR,NL,SF1,H,HO,DELTA,ERROR,E,ITER,WBE,
     *ISTEP,Q,R,NIC,RS,MAS,SF2,QDY,MDY,CH,CELE,QF,QRD,QST,
     *MDR,QDRAIN,QE,RH,RD,T1,T2,QR,QP,MSB,MF1,MDM1,MS1,MD1,
     *ME1,QI,MFW,MUS,HSB,MAP1,IMP,JMP,MAU,BOT)
C
C     THIS SUBROUTINE CALCULATES AND DISPLAYS WATER
C     BALANCES
C
C     METACOMMANDS
C
$LARGE
C
C     DECLARATIONS
C
      DIMENSION STORE(4),PUMP(4),ALU(4),ALD(4),SF1(20,20,4),H(20,20,4),
     *HO(20,20,4),DELTA(50),DIFF(4),Q(20,20,4),R(20,20,4),TOTIN(4),
     *TOTOUT(4),RS(20,20,4),STORE1(4),QDY(20,20,1),SF2(20,20,4),
     *RINJ(4),RRECH(1),FWELL(4),DMINE(4),SRECH(1),DPUMP(1),EVAPO(1),
     *QWELL(4),CH(20,20,4),CELE(20,20,4),QF(20,20,4),QRD(20,20,4),
     *QST(20,20,1),MDR(20,20,1),QDRAIN(20,20,1),QE(20,20,1),
     *RH(20,20,1),RD(20,20,1),T1(20,20,4),T2(20,20,4),QR(20,20,1),
     *QP(20,20,4),MSB(20,20,1),QI(20,20,4),MFW(20,20,4),BOT(20,20,4)
C
C     INITIALIZE VARIABLES
C
      DO 8 K=1,NL
      STORE(K)=0.0
      STORE1(K)=0.0
      PUMP(K)=0.0
      ALU(K)=0.0
      ALD(K)=0.0
      RINJ(K)=0.0
      IF(K.EQ.1)RRECH(K)=0.0
      FWELL(K)=0.0
      DMINE(K)=0.0
      IF(K.EQ.1)SRECH(1)=0.0
      IF(K.EQ.1)DPUMP(1)=0.0
      IF(K.EQ.1)EVAPO(1)=0.0
      QWELL(K)=0.0
    8 CONTINUE
C
C     CALCULATE WATER BALANCES
C
      DO 40 K=1,NL
      DO 30 J=1,NR
```

```
      DO 21 I=1,NC
      QRD(I,J,K)=0.0
      S=SF1(I,J,K)
      IF(MDY.EQ.1)S=SF2(I,J,K)
      IF(H(I,J,K).LT.CH(I,J,K))S=SF2(I,J,K)
      STORE(K)=STORE(K)+S*ABS(HO(I,J,K)-H(I,J,K))/DELTA(ISTEP)
      IF((HO(I,J,K)-CH(I,J,K))*(H(I,J,K)-CH(I,J,K)).LT.0.0)
     *STORE(K)=STORE(K)+ABS(HO(I,J,K)-CH(I,J,K))*(SF1(I,J,K)
     *-SF2(I,J,K))/DELTA(ISTEP)
      IF(MDY.EQ.1.AND.K.EQ.1)STORE(K)=STORE(K)+QDY(I,J,K)
      IF(K.EQ.1.AND.MUS.EQ.2)GOTO 10
      IF(MAU.EQ.1.OR.MAU.EQ.2.OR.MAU.EQ.3)GOTO 9
      STORE1(K)=STORE1(K)+RS(I,J,K)*ABS(HO(I,J,K)-H(I,J,K))
    9 PUMP(K)=PUMP(K)+Q(I,J,K)
      IF(K.EQ.1.AND.QE(I,J,1).GT.0.0)PUMP(K)=PUMP(K)-QE(I,J,1)
      RINJ(K)=RINJ(K)+QI(I,J,K)
      IF(K.EQ.1)RRECH(K)=RRECH(K)+QR(I,J,K)
      QWELL(K)=QWELL(K)+QP(I,J,K)
      IF(MAP1.EQ.1)QWELL(K)=QWELL(K)+Q(IMP,JMP,K)
      IF(MFW(I,J,K).NE.1)GOTO 10
      QF(I,J,K)=R(I,J,K)*(H(I,J,K)-CELE(I,J,K))
      IF(H(I,J,K).LE.CELE(I,J,K))QF(I,J,K)=0.0
      FWELL(K)=FWELL(K)+QF(I,J,K)
   10 IF(SF1(I,J,K).LT.1E21.OR.SF2(I,J,K).LT.1E21)GOTO 15
      IF((J+1).GE.NR)GOTO 11
      IF(SF1(I,J+1,K).EQ.1E21.OR.SF2(I,J+1,K).EQ.1E21)GOTO 11
      QRD(I,J,K)=T1(I,J,K)*ABS(H(I,J+1,K)-H(I,J,K))
   11 IF((I+1).GE.NC)GOTO 12
      IF(SF1(I+1,J,K).EQ.1E21.OR.SF2(I+1,J,K).EQ.1E21)GOTO 12
      QRD(I,J,K)=QRD(I,J,K)+T2(I,J,K)*ABS(H(I+1,J,K)-H(I,J,K))
   12 IF((J-1).LE.1.)GOTO 13
      IF(SF1(I,J-1,K).EQ.1E21.OR.SF2(I,J-1,K).EQ.1E21)GOTO 13
      QRD(I,J,K)=QRD(I,J,K)+T1(I,J-1,K)*ABS(H(I,J-1,K)-H(I,J,K))
   13 IF((I-1).LE.1.)GOTO 14
      IF(SF1(I-1,J,K).EQ.1E21.OR.SF2(I-1,J,K).EQ.1E21)GOTO 14
      QRD(I,J,K)=QRD(I,J,K)+T2(I-1,J,K)*ABS(H(I-1,J,K)-H(I,J,K))
   14 DMINE(K)=DMINE(K)+QRD(I,J,K)
   15 IF(K.EQ.1)THEN
      IF(MSB(I,J,1).EQ.1)THEN
      IF(H(I,J,1).LE.RD(I,J,1))THEN
      QST(I,J,1)=R(I,J,1)*(RH(I,J,1)-RD(I,J,1))
      SRECH(K)=SRECH(K)+QST(I,J,1)
      ENDIF
      QST(I,J,1)=R(I,J,1)*(RH(I,J,1)-H(I,J,1))
      SRECH(K)=SRECH(K)+QST(I,J,1)
      ENDIF
      IF(MDR(I,J,1).EQ.1)THEN
      QDRAIN(I,J,1)=ABS(R(I,J,1)*(RH(I,J,1)-H(I,J,1)))
      IF(H(I,J,1).LE.RH(I,J,1))THEN
      QDRAIN(I,J,1)=0.0
      ENDIF
      DPUMP(K)=DPUMP(K)+QDRAIN(I,J,1)
```

```
      ENDIF
      IF(QE(I,J,1).GT.0.0.AND.H(I,J,1).GT.RD(I,J,1))
     *EVAPO(K)=EVAPO(K)+QE(I,J,1)-(R(I,J,1)*(RH(I,J,1)-H(I,J,1)))
      IF(QE(I,J,1).GT.0.0.AND.H(I,J,1).LE.RD(I,J,1))
     *EVAPO(K)=0.0
      ENDIF
   16 IF(K.EQ.1.AND.K-NL.EQ.0)THEN
      ALU(K)=ALU(K)+0.0
      IF(MUS.EQ.2)ALD(K)=ALD(K)+0.0
      IF(MUS.EQ.1)ALD(K)=ALD(K)+(HSB-H(I,J,K))*R(I,J,K)
      IF(MUS.EQ.1.AND.H(I,J,K).LT.CH(I,J,K))ALD(K)=ALD(K)+
     *(HSB-CH(I,J,K))*R(I,J,K)
      GOTO 21
      ENDIF
      IF(K.EQ.1.AND.K-NL.NE.0)THEN
      ALU(K)=(H(I,J,K+1)-H(I,J,K))*R(I,J,K)
      IF(H(I,J,K+1).LT.CH(I,J,K+1))ALU(K)=(CH(I,J,K+1)-H(I,J,K))*
     *R(I,J,K)
      IF(MUS.EQ.2)ALD(K)=ALD(K)+0.0
      IF(MUS.EQ.1)ALD(K)=ALD(K)+(HSB-H(I,J,K))*R(I,J,K)
      IF(MUS.EQ.1.AND.H(I,J,K).LT.CH(I,J,K))ALD(K)=ALD(K)+
     *(HSB-CH(I,J,K))*R(I,J,K)
      GOTO 21
      ENDIF
      IF(K-NL.EQ.0)THEN
      ALU(K)=ALU(K)+0.0
      ALD(K)=ALD(K)+(H(I,J,NL-1)-H(I,J,NL))*R(I,J,NL)
      IF(H(I,J,NL).LT.CH(I,J,NL))
     *ALD(K)=ALD(K)+(H(I,J,NL-1)-CH(I,J,NL))*R(I,J,NL)
      GOTO 21
      ENDIF
      ALU(K)=ALU(K)+(H(I,J,K+1)-H(I,J,K))*R(I,J,K)
      ALD(K)=ALD(K)+(H(I,J,K-1)-H(I,J,K))*R(I,J,K)
      IF(H(I,J,K).LT.CH(I,J,K))
     *ALD(K)=ALD(K)+(H(I,J,K-1)-CH(I,J,K))*R(I,J,K)
   21 CONTINUE
   30 CONTINUE
   40 CONTINUE
      DO 50 K=1,NL
      TOTIN(K)=STORE(K)+STORE1(K)+ALD(K)+
     *ALU(K)+RINJ(K)
      IF(K.EQ.1)TOTIN(K)=TOTIN(K)+SRECH(K)+RRECH(K)
      TOTOUT(K)=PUMP(K)+FWELL(K)+DMINE(K)
      IF(K.EQ.1)TOTOUT(K)=TOTOUT(K)+DPUMP(K)+EVAPO(K)
      IF(TOTOUT(K).EQ.0.0.OR.TOTIN(K).EQ.0.0)THEN
      DIFF(K)=0.0
      GOTO 50
      ENDIF
      DIFF(K)=ABS(1.0-TOTOUT(K)/TOTIN(K))*100.0
   50 CONTINUE
C
C     PRINT INTERMEDIATE RESULTS
C
```

```
      IF(NIC.NE.1)GOTO 240
      WRITE(*,80)
 80   FORMAT('0','PRELIMINARY COMPUTATION RESULTS:')
      WRITE(*,90)ISTEP
 90   FORMAT('0','Time increment number=',I3)
      WRITE(*,100)ITER
100   FORMAT(1X,'Iteration number=',I4)
112   WRITE(*,120)ERROR,E
120   FORMAT(1X,'Head error in ft (should be <',F9.2,/1X,
     *')',' and is ',F9.2)
      DO 238 K=1,NL
      WRITE(*,122)K
122   FORMAT(1X,'Unit or layer number=',I2)
      IF(K-NL.EQ.0)WRITE(*,124)WBE,DIFF(K)
124   FORMAT(1X,'Water balance error in percent (should be <',/1X,
     *F6.2,')',' and is ',F6.2)
126   WRITE(*,130)STORE(K)
130   FORMAT(1X,'Component of flow from aquifer'/1X,
     *'storage in gpd=',F14.2)
      WRITE(*,150)ALD(K)
150   FORMAT(1X,'Component of flow from leakage'/1X,
     *'above aquifer or between aquifer layers in gpd=',F14.2)
      WRITE(*,160)ALU(K)
160   FORMAT(1X,'Component of flow from leakage'/1X,
     *'below aquifer or between aquifer layers in gpd=',F12.2)
      IF(MAS.EQ.1)WRITE(*,162)STORE1(K)
162   FORMAT(1X,'Component of flow from aquitard'/1X,
     *'storage in gpd=',F12.2)
      WRITE(*,164)QWELL(K)
164   FORMAT(1X,'Component of flow from discharge'/1X,
     *'units in gpd=',F14.2)
      IF(RINJ(K).GT.0.0)WRITE(*,166)RINJ(K)
166   FORMAT(1X,'Component of flow from injection'/1X,
     *'well recharge in gpd=',F14.2)
      IF(K.EQ.1.AND.RRECH(K).GT.0.0)WRITE(*,168)RRECH(K)
168   FORMAT(1X,'Component of flow from areal'/1X,
     *'recharge in gpd=',F12.2)
      IF(MF1.EQ.1)WRITE(*,170)FWELL(K)
170   FORMAT(1X,'Flowing well discharge in gpd=',F14.2)
      IF(MDM1.EQ.1)WRITE(*,172)DMINE(K)
172   FORMAT(1X,'Component of flow from mine'/1X,
     *'drainage in gpd=',F14.2)
      IF(MS1.EQ.1.AND.K.EQ.1)WRITE(*,174)SRECH(K)
174   FORMAT(1X,'Component of flow from induced'/1X,
     *'infiltration in gpd=',F14.2)
      IF(MD1.EQ.1.AND.K.EQ.1)WRITE(*,176)DPUMP(K)
176   FORMAT(1X,'Component of flow from drains in gpd=',F14.2)
      IF(ME1.EQ.1.AND.K.EQ.1)WRITE(*,178)EVAPO(K)
178   FORMAT(1X,'Component of flow from evapotranspiration'/1X,
     *'in gpd=',F14.2)
238   CONTINUE
      PAUSE
240   END
```

```
C
      SUBROUTINE WELSTOR(ISTEP,QAT)
C
C     THIS SUBROUTINE SIMULATES WELL STORAGE CAPACITY
C     BY SUCCESSIVE APPROXIMATIONS OF DISCHARGE FROM
C     AQUIFER
C
C     METACOMMANDS
C
$LARGE
C
   10 WRITE(*,20)
   20 FORMAT('0','Enter 1 to simulate well storage capacity'/1X,
     *'or 2 for no well storage capacity. The time after'/1X,
     *'pumping started in minutes beyond which well storage'/1X,
     *'capacity impacts are negligible is equal to 540000'/1X,
     *'x (production well radius in ft squared minus pump'/1X,
     *'column pipe radius in ft squared) divided by aquifer'/1X,
     *'transmissivity in gpd/ft  '\)
      READ(*,*)MST
      IF(MST.NE.1.AND.MST.NE.2)GOTO 10
      IF(MST.EQ.2)GOTO 250
   25 IF(ISTEP.EQ.1)THEN
      QAT1=0
      ENDIF
C
C     DATA BASE SPECIFICATION
C
   30 WRITE(*,40)
   40 FORMAT('0','It is assumed that only one production'/1X,
     *'well in the model has well storage capacity'/)
      WRITE(*,50)
   50 FORMAT(1X,'Enter real number drawdown (positive) at'/1X,
     *'production well in ft during present time increment'/1X,
     *'based on earlier calculations of drawdown with'/1X,
     *'well storage capacity at beginning of present time'/1X,
     *'increment and drawdown without well storage at end'/1X,
     *'of present time increment. Should include any partial'/1X,
     *'penetration and well loss impacts.  '\)
      READ(*,*)DDT
      WRITE(*,52)
   52 FORMAT(1X,'Enter real number drawdown at product-'/1X,
     *'ion well in ft at beginning of present time increment'/1X,
     *'with well storage capacity based on earlier calculat-'/1X,
     *'ions (must be positive number)  '\)
      READ(*,*)DDT1
```

```
      WRITE(*,60)
   60 FORMAT(1X,'Enter real number production well radius'/1X,
     *'in ft  '\)
      READ(*,*)RW
      WRITE(*,70)
   70 FORMAT(1X,'Enter real number discharge pipe-column'/1X,
     *'radius in ft  '\)
      READ(*,*)RC
      WRITE(*,80)
   80 FORMAT(1X,'Enter real number production well discharge'/1X,
     *'rate in gpm  '\)
      READ(*,*)Q
      IF(ISTEP.GT.1)THEN
      QAT1=QAT
      ENDIF
      WRITE(*,90)
   90 FORMAT(1X,'Enter real number time at end of present'/1X,
     *'simulation period in minutes  '\)
      READ(*,*)DEL
   95 WRITE(*,100)
  100 FORMAT('0','Enter 1 to calculated drawdown at production'/1X,
     *'well or 2 to calculate drawdown at observation well  '\)
      READ(*,*)MS1
      IF(MS1.NE.1.AND.MS1.NE.2)GOTO 95
      IF(MS1.EQ.1)GOTO 120
      WRITE(*,110)
  110 FORMAT('0','Enter real number drawdown (positive) at'/1X,
     *'observation well in ft during present time increment'/1X,
     *'based on earlier calculations of drawdown at begin-'/1X,
     *'ning of present time increment with well storage'/1X,
     *'capacity and drawdown at end of time increment'/1X,
     *'without well storage capacity. Should include any'/1X,
     *'partial penetration impact  '\)
      READ(*,*)DDD
      WRITE(*,112)
  112 FORMAT(1X,'Enter real number drawdown at observat-'/1X,
     *'ion well in ft at beginning of present time increment'/1X,
     *'with well storage capacity based on earlier calculat-'/1X,
     *'ions (must be positive number)  '\)
      READ(*,*)DDD1
  120 S1=DDT
      D1=0
      D2=S1
  122 CS=(D1+D2)/2
  130 QAT=2*Q-(2*3.1416*(RW**2-RC**2)*CS*7.48/DEL)-QAT1
      D3=(QAT+QAT1)*S1/(2*Q)
      IF(ABS(CS-D3).LT.0.001)GOTO 180
      IF(D3.LT.CS)D2=CS
      IF(D3.GT.CS)D1=CS
      GOTO 122
  180 IF(MS1.EQ.1)DDT=DDT1+DDT*(QAT+QAT1)/(2*Q)
      IF(MS1.EQ.2)DDD=DDD1+DDD*(QAT+QAT1)/(2*Q)
```

```
200 IF(MS1.EQ.1)WRITE(*,210)DDT
210 FORMAT('0','Drawdown at production well in ft with'/1X,
   *'well storage capacity=',F9.2)
    IF(MS1.EQ.2)WRITE(*,220)DDD
220 FORMAT('0','Drawdown at observation well in ft with'/1X,
   *'well storage capacity=',F9.2)
230 WRITE(*,240)
240 FORMAT('0','Enter 1 for another well storage capacity'/1X,
   *'calculation 2 to continue program  '\)
    READ(*,*)MS2
    IF(MS2.NE.1.AND.MS2.NE.2)GOTO 230
    IF(MS2.EQ.1)GOTO 25
250 END
```

```
C
      SUBROUTINE XYZDT(H,NC,NR,DELX,DELY)
C
C     THIS SUBROUTINE CONVERTS HEAD ARRAYS
C     INTO XYZ DATA FOR USE IN GRAPHICS PROGRAMS
C
C     METACOMMANDS
C
$LARGE
C
C     DECLARATIONS
C
      CHARACTER*14FNAME
      DIMENSION H(20,20,4),XSUM(20),YSUM(20),
     *DELX(21),DELY(21)
C
   10 WRITE(*,20)
   20 FORMAT('0','Enter 1 to convert head data to a standard'/1X,
     *'ASCII file containing one x-coordinate (X), one'/1X,
     *'y-coordinate (Y), and one head coordinate (Z)'/1X,
     *'entry per line; the (X,Y,Z) values on a single line'/1X,
     *'are separated by a comma; or 2 for no conversion file  '\)
      READ(*,*)NZ
      IF(NZ.NE.1.AND.NZ.NE.2)GOTO 10
   21 IF(NZ.EQ.1)THEN
      WRITE(*,22)
   22 FORMAT('0','Enter integer unit or layer number  '\)
      READ(*,*)M
      DO 30 J=1,NR
      IF(J.EQ.1)THEN
      YSUM(J)=0.5*DELY(J)
      GOTO 30
      ENDIF
      YSUM(J)=YSUM(J-1)+0.5*DELY(J-1)+0.5*DELY(J)
   30 CONTINUE
      DO 40 I=1,NC
      IF(I.EQ.1)THEN
      XSUM(I)=0.5*DELX(I)
      GOTO 40
      ENDIF
      XSUM(I)=XSUM(I-1)+0.5*DELX(I-1)+0.5*DELX(I)
   40 CONTINUE
      WRITE(*,70)
   70 FORMAT('0','Enter a valid X,Y,Z data base specification'/1X,
     *'filename with extension DAT of ASCII file to be exported'/1X,
     *'such as B:FILE6.DAT in apostrophes. If apostrophes'/1X,
```

```
     *'are ommitted, a fatal error will occur.  '\)
      READ(*,*)FNAME
      OPEN(6,FILE=FNAME,STATUS='NEW')
      DO 100 J=1,NR
      DO 90 I=1,NC
      WRITE(6,80)XSUM(I),YSUM(J),H(I,J,M)
  80  FORMAT(1X,E12.4,',',E12.4,',',E12.4)
  90  CONTINUE
 100  CONTINUE
      CLOSE (6)
 110  WRITE(*,120)
 120  FORMAT('0','Enter 1 to create a file for heads'/1X,
     *'in another unit or layer or 2 to end'/1X,
     *'program  '\)
      READ(*,*)N1H
      IF(N1H.NE.1.AND.N1H.NE.2)GOTO 110
      IF(N1H.EQ.1)GOTO 21
      ENDIF
      END
```

Appendix B
Source Code for GWTR3D

The source code, created with the Line Editor Edlin included in the MS-Disk Operating System (trademark of Microsoft Corp.), for GWTR3D is listed in this appendix. Instructions for using Edlin to create and edit files are provided in the MS-Disk Operating System manual and by Wolverton (1984, pp. 196–231). The program is written in FORTRAN 77 (see Etter, 1987 and Fouri, et al., 1986) specifically for the Microsoft FORTRAN Compiler version 3.2 (trademark of Microsoft Corp.) and an IBM PC (trademark of International Business Machines Corp.). Program algorithms are documented as COMMENT statements in lists to assist the reader in understanding and modifying the codes. Many statements in the programs could be restructured or eliminated to produce more efficient coding, but they have been intentionally written in their present form to provide clear explanations of processing steps. GWTR3D consists of a main program and 34 subroutines filling 118115 bytes of memory. The number of subroutines and functions was made large in an effort to make the program understandable to inexperienced users and to enable the use of EDLIN which is designed for short files. The experienced user may wish to consolidate some of the subroutines and functions with a more sophisticated word processor.

GWTR3D SUBROUTINES

ADD
ADVAN
ANORM
AREA3D
ARESNK
ARESOR
CLEAR
CONMAP
CONSTR
CONTIN
DATPRN
DISPER
FILEI1
FILEI2
FILEI3
FILE01
FILE02
FILE03
HEAD
INFOM
INIT
INJIN
LOCSOR
MAP
MOVE
PRESOR
PROFIL
RAN
SLUGIN
SNKCON
SUMMRY
VELO
VIJK
XYZDTC

```
      PROGRAM GWTR3D
C
C      METACOMMANDS
C
$LARGE
C
C      DECLARATIONS
C
      CHARACTER*14FNAME
      DIMENSION CSINK(50,50),CSLUG(50,50),XSUM(50),YSUM(50),
     *HHE(50,50),CNODE(50,10)
      COMMON/NSUB1/NC,NR
      COMMON/NSUB2/NPUMP,NPITS
      COMMON/NSUB3/IJJ(24),JIJ(24)
      COMMON/NSUB4/IP(24),JP(24),ISS(100),JSS(100)
      COMMON/NSUB6/MAXP,NSORS,NIJ,NSSN,MARK(50,50),NRAD
      COMMON/NSUB7/NP,NPART(50,50)
      COMMON/NSUB8/NAL,NLSK,NTT,NPE,NN1,IWS,NLC,NI1
      COMMON/NSUB9/LEK,NRE,NCS
      COMMON/ASUB1/CONSNK(50,50),CONC(50,50),
     *CSCC(50,50),CONSOR(50,50)
      COMMON/ASUB2/CH(50,50),BOT(50,50),H(50,50)
      COMMON/ASUB3/P3(50,50),P4(50,50),P5(50,50),THICK3(50,50)
      COMMON/ASUB4/THICK4(50,50),THICK5(50,50),H1(50,50),H2(50,50)
      COMMON/AREA1/RH(50,05),R(50,50),RD(50,50),RS(50,50)
      COMMON/AREA2/DELX(51),DELY(51),P1(50,50),P2(50,50)
      COMMON/AREA3/RD1,HL,PLEAK(50),Q(50,50),RD1V
      COMMON/AREA5/X(5001),Y(5001),V(50,50,2)
      COMMON/AREA6/DELTA,THICKC(50,50),PS(50,50)
      COMMON/AREA7/PM,DISPL,DISPT,EPOR,APOR
      COMMON/AREA8/TMAP,ANC,ANR,TABLE(124)
C
C      PROGRAM INFORMATION
C
      CALL INFOM
   80 WRITE(*,85)
   85 FORMAT('0','Enter 1 for information display at intervals'/1X,
     *'during program or 2 for no information display  '\)
      READ(*,*)LY1
      IF(LY1.NE.1.AND.LY1.NE.2)GOTO 80
      IF(LY1.EQ.2)GOTO 109
      WRITE(*,100)
  100 FORMAT('0','This program simulates contaminant transport'/1X,
     *'from multiple slug and continuous sources towards sinks'/1X,
     *'in one aquifer or aquifer layer during one time increment'/1X,
     *'with or without pre-existing conditions. Vertical'/1X,
```

```
     *'contaminant transport is simulated through areal sink and'/1X,
     *'source subroutines. For example, during program run 1'/1X,
     *'vertical leakage of contaminants from aquifer 1 through'/1X,
     *'an aquitard and into aquifer 2 may be simulated as areal'/1X,
     *'sink mass. During program run 2, this areal sink mass'/1X,
     *'becomes areal source mass in aquifer 2. Thus, 3-D contam-'/1X,
     *'inant transport during multiple time increments may be'/1X,
     *'simulated with multiple program runs. Only English gallon-'/1X,
     *'day-ft system of units are supported for groundwater flow.'/1X,
     *'Milligram-liter-pound units are supported for groundwater'/1X,
     *'transport. Units are specified in interactive data base'/1X,
     *'input statements. If your units do not match those units,'/1X,
     *'then use conversion tables to obtain specified units.')
      PAUSE
      WRITE(*,105)
  105 FORMAT('0','Contaminant mass is assigned to a large number'/1X,
     *'of particles with x,y coordinates. Particle coordinates are'/1X,
     *'advanced based on time dependent head maps consistent with'/1X,
     *'advective migration controlled by groundwater velocities'/1X,
     *'and dispersive migration simulated by random particle move-'/1X,
     *'ments. Particles within 1/2 grid spacing of a sink are'/1X,
     *'captured by the sink. Other particles are assigned to the'/1X,
     *'nearest node at end of time increment. The number of'/1X,
     *'particles is translated into contaminant concentration.'/1X,
     *'Sources and sinks are assumed to fully penetrate the'/1X,
     *'aquifer. Contaminant concentration is assumed not to affect'/1X,
     *'the liquid density and viscosity. Sorption and radioactive'/1X,
     *'decay may be simulated.')
      PAUSE
C
C     INTERACTIVE DATA BASE INPUT
C
  109 WRITE(*,110)
  110 FORMAT('0','MODEL DATA BASE SPECIFICATION:')
      WRITE(*,120)
  120 FORMAT('0','Please be careful in responding to interactive'/1X,
     *'data base prompts. Option prompts will be repeated if you'/1X,
     *'respond incorrectly. Be sure you enter either a real or'/1X,
     *'integer number as specified in prompt statements. If you'/1X,
     *'enter a real number in response to an option choice prompt'/1X,
     *'instead of an integer, an error will occur. If you enter'/1X,
     *'an invalid filename when prompted or the filename is'/1X,
     *'not enclosed in apostrophes, an error will occur.')
      WRITE(*,130)
  130 FORMAT('0','Program requires printer, turn it on or else'/1X,
     *'an error will occur. After data base input is complete'/1X,
     *'calculations will start as indicated by flashing prompt'/1X,
     *'symbol on left side of screen. Be patient, calculations may'/1X,
     *'take minutes to hours.')
  140 WRITE(*,150)
  150 FORMAT('0','Enter 1 for keyboard monitor well, sink, grid,'/1X,
     *'hydraulic characteristic data base input or 2 to'/1X,
```

```
     *'import ASCII data base file previously created with this'/1X,
     *'program. Formats of imported and exported data base'/1X,
     *'files must match exactly. Attempts to import files with'/1X,
     *'improper formats or an illegal filename will result in a'/1X,
     *'fatal error requiring that diskettes be removed from'/1X,
     *'drives, the computer shut off, and the program re-loaded'/1X,
     *'and run.  '\)
      READ(*,*)KFI
      IF(KFI.NE.1.AND.KFI.NE.2)GOTO 140
      IF(KFI.EQ.2)THEN
C     IMPORT DATA BASE
      CALL FILEI1(NN8)
      OPEN(4,FILE='LPT1:')
      ENDIF
  152 WRITE(*,153)
  153 FORMAT('0','Enter 1 if pre-existing contaminant con-'/1X,
     *'centration is appreciable. The model grid and any pre-'/1X,
     *'existing contaminant loads must extend sufficiently'/1X,
     *'beyond the area under investigation so that migration'/1X,
     *'of pre-existing particles to and from that area is'/1X,
     *'simulated. Enter 2 for no pre-existing contaminant'/1X,
     *'concentration.  '\)
      READ(*,*)NPE
      IF(NPE.NE.1.AND.NPE.NE.2)GOTO 152
      IF(NPE.EQ.1.AND.KFI.EQ.2)CALL FILEI3
  160 IF(KFI.EQ.2)GOTO 2261
C
C     DEFINE OUTPUT DEVICE NUMBER
C
      OPEN(4,FILE='LPT1:')
C
      WRITE(4,350)
  350 FORMAT(1X,'MODEL DATA BASE SPECIFICATION:')
C
C     DEFINE DEFAULT MODEL GRID
C
      IF(LY1.EQ.2)GOTO 365
      WRITE(*,360)
  360 FORMAT('0','The model grid lines superposed over the'/1X,
     *'aquifer are indexed using the I (column), J (row)'/1X,
     *'notation colinear with the X and Y directions,'/1X,
     *'respectively. I-coordinates increase left to'/1X,
     *'right and J-coordinates increase top to bottom. Grid'/1X,
     *'spacing is centered at the nodes (grid line inter-'/1X,
     *'sections). The origin is the upper-left corner.'/1X,
     *'Contaminants migrate on the basis of an X,Y coordinate'/1X,
     *'system congruent with and on the same scale as the I,J'/1X,
     *'grid. Grid boundaries must extend beyond contaminant'/1X,
     *'plume. The number of grid columns and rows must match'/1X,
     *'the number of grid columns and rows in head data base')
  365 WRITE(*,410)
  410 FORMAT('0','Enter integer number of grid columns'/1X,
```

```
     *'(must be <51)  '\)
      READ(*,*)NC
      WRITE(4,420)NC
  420 FORMAT('0','Number of grid columns=',I3)
      WRITE(*,430)
  430 FORMAT(1X,'Enter integer number of grid rows'/1X,
     *'(must be <51)  '\)
      READ(*,*)NR
      WRITE(4,440)NR
  440 FORMAT(1X,'Number of grid rows=',I3)
      WRITE(*,450)
  450 FORMAT('0','The default grid setting is uniform (square).'/1X,
     *'Enter the real number default grid spacing in ft,'/1X,
     *'usually ranges from 10-10000 ft  '\)
      READ(*,*)GS
      WRITE(4,460)GS
  460 FORMAT(1X,'Default (square) grid spacing in ft=',F6.0)
      DO 470 I=1,NC+1
      DELX(I)=GS
  470 CONTINUE
      DO 480 J=1,NR+1
      DELY(J)=GS
  480 CONTINUE
  490 WRITE(*,500)
  500 FORMAT('0','Enter 1 for a variable grid or 2'/1X,
     *'to retain default grid setting'/1X,
     *'(a uniform grid is desirable).  '\)
      READ(*,*)NVGS
      IF(NVGS.NE.1.AND.NVGS.NE.2) GOTO 490
      IF(NVGS.EQ.2) THEN
      WRITE(4,510)
  510 FORMAT(1X,'Grid setting is uniform (square)')
      GOTO 699
      ENDIF
C
C     DEFINE VARIABLE GRID SPACING
C
  520 WRITE(4,530)
  530 FORMAT(1X,'Grid is variable'/)
  540 WRITE(*,550)
  550 FORMAT('0','Enter integer number of grid blocks to'/1X,
     *'be changed to variable grid settings  '\)
      READ(*,*)NGB
      DO 621 K=1,NGB
      WRITE(*,560)K
  560 FORMAT('0','Grid block number',I3)
      WRITE(*,570)
  570 FORMAT(1X,'Enter integer grid block left- and right-'/1X,
     *'most column I coordinates separated by a comma  '\)
      READ(*,*)IVG1,IVG2
      WRITE(*,580)
  580 FORMAT(1X,'Enter integer grid block upper- and lower-'/1X,
```

```
     *'row J coordinates separated by a comma  '\)
      READ(*,*)JVG1,JVG2
      DO 600 I=IVG1,IVG2
      WRITE(*,590)I
  590 FORMAT(1X,'Enter real number grid spacing in ft'/1X,
     *'starting at column',I3,' (adjoining grid spacings'/1X,
     *'should be limited to double or less)  '\)
      READ(*,*)DELX(I)
  600 CONTINUE
      DO 620 J=JVG1,JVG2
      WRITE(*,610)J
  610 FORMAT(1X,'Enter real number grid spacing in ft'/1X,
     *'starting at row',I3,' (adjoining grid spacings'/1X,
     *'should be limited to double or less)  '\)
      READ(*,*)DELY(J)
  620 CONTINUE
  621 CONTINUE
  622 WRITE(*,624)
  624 FORMAT('0','Enter 1 for hard copy of grid spacings'/1X,
     *'or 2 to continue program  '\)
      READ(*,*)MH1
      IF(MH1.NE.1.AND.MH1.NE.2)GOTO 622
      IF (MH1.EQ.2)GOTO 692
      WRITE(4,630)
  630 FORMAT('0',5X,'Column Number',3X,'Grid Spacing (ft)')
      DO 650 I=1,NC
      WRITE(4,640)I,DELX(I)
  640 FORMAT(1X,8X,I3,9X,F9.2)
  650 CONTINUE
      WRITE(4,660)
  660 FORMAT('0',5X,'Row Number',6X,'Grid Spacing (ft)')
      DO 680 J=1 ,NR
      WRITE(4,670)J,DELY(J)
  670 FORMAT(1X,8X,I3,9X,F9.2)
  680 CONTINUE
  692 WRITE(*,694)
  694 FORMAT('0','Enter 1 to revise variable grid data base or'/1X,
     *'2 to continue data base specification  '\)
      READ(*,*)NRGB
      IF(NRGB.NE.1.AND.NRGB.NE.2) GOTO 692
      IF(NRGB.EQ.1) GOTO 520
C
C     DEFINE DURATION OF TIME INCREMENT
C
  699 WRITE(*,700)
  700 FORMAT('0','If areal leakage sink sources are simulated,'/1X,
     *'duration of time increment should be the same as that used'/1X,
     *'in conjunction with the generation of areal leakage sinks')
  702 WRITE(*,704)
  704 FORMAT('0','Enter a real number (such as 10.0) value for'/1X,
     *'duration of time increment in days  '\)
      READ(*,*)DELTA
```

```
      WRITE(4,706)DELTA
  706 FORMAT(1X,'Duration of time increment in days=',F10.3)
     *'in days=',F10.3)
C
C     FILL ARRAYS WITH AQUIFER OR AQUIFER LAYER
C     DEFAULT HYDRAULIC CHARACTERISTICS AND GEOMETRY
C     VALUES
C
  720 WRITE(*,730)
  730 FORMAT('0','After each of the following aquifer'/1X,
     *'hydraulic characteristics and geometries is displayed'/1X,
     *'enter the real number default value. Elevations refer'/1X,
     *'to mean sea level or zero base level. If a zero base'/1X,
     *'level is used, some elevations will be negative.')
      WRITE(*,740)
  740 FORMAT('0','Aquifer or aquifer layer hydraulic conduct-'/1X,
     *'ivity in gpd/sq ft, usually ranges from 1 to'/1X,
     *'10000 gpd/sq ft  '\)
      READ(*,*)PP
      WRITE(4,750)PP
  750 FORMAT(1X,'Default aquifer or aquifer layer hydraulic'/1X,
     *'conductivity in gpd/sq ft=',F9.2)
      WRITE(*,760)
  760 FORMAT(1X,'Top of aquifer or aquifer layer elevation in ft  '\)
      READ(*,*)CHH
      WRITE(4,770)CHH
  770 FORMAT(1X,'Default top of aquifer or aquifer layer elevation'/1X,
     *'in ft=',F9.2)
      WRITE(*,780)
  780 FORMAT(1X,'Base of aquifer or aquifer layer elevation'/1X,
     *'in ft  '\)
      READ(*,*)BOTT
      WRITE(4,790)BOTT
  790 FORMAT(1X,'Default base of aquifer or aquifer layer'/1X,
     *'elevation in ft=',F9.2)
      DO 830 J=1,NR
      DO 820 I=1,NC
      P1(I,J)=PP
      P2(I,J)=PP
      CH(I,J)=CHH
      BOT(I,J)=BOTT
C
C     INITIALIZE VARIABLES
C
      THICKC(I,J)=0.0
      RH(I,J)=0.0
      RD(I,J)=0.0
      R(I,J)=0.0
      CONSOR(I,J)=0.0
      RS(I,J)=0.0
      CSCC(I,J)=0.0
      V(I,J,1)=0.0
```

```
      V(I,J,2)=0.0
      PS(I,J)=0.0
      P3(I,J)=0.0
      P4(I,J)=0.0
      P5(I,J)=0.0
      THICK3(I,J)=0.0
      THICK4(I,J)=0.0
      THICK5(I,J)=0.0
  820 CONTINUE
  830 CONTINUE
C
C     DEFINE AQUIFER POROSITY
C
      WRITE(*,831)
  831 FORMAT('0','Aquifer or aquifer layer porosity is'/1X,
     *'assumed to be uniform in space and time')
      WRITE(*,832)
  832 FORMAT(1X,'Enter real number actual aquifer or aquifer'/1X,
     *'layer porosity as a decimal, usually ranges from 0.05'/1X,
     *'to .50  '\)
      READ(*,*)APOR
      WRITE(4,833)APOR
  833 FORMAT(1X,'Actual aquifer or aquifer layer porosity'/1X,
     *'as a decimal=',F7.4)
      WRITE(*,834)
  834 FORMAT(1X,'Enter real number effective aquifer or aquifer'/1X,
     *'layer porosity as a decimal usually ranges from .04'/1X,
     *'to .4  '\)
      READ(*,*)EPOR
      WRITE(4,836)EPOR
  836 FORMAT(1X,'Effective aquifer or aquifer layer porosity'/1X,
     *'as a decimal=',F7.4)
  837 WRITE(*,838)
  838 FORMAT('0','Enter 1 to revise default values or'/1X,
     *'2 to continue program  '\)
      READ(*,*)NR1
      IF(NR1.NE.1.AND.NR1.NE.2)GOTO 837
      IF(NR1.EQ.1)GOTO 720
      CALL INIT
  840 WRITE(*,850)
  850 FORMAT('0','Enter 1 to change default values at'/1X,
     *'selected nodes or 2 to retain default values  '\)
      READ(*,*)NDV
      IF(NDV.NE.1.AND.NDV.NE.2) GOTO 840
      IF(NDV.EQ.2)GOTO 1040
C
C     DEFINE NODAL HYDRAULIC CHARACTERISTIC
C     AND AQUIFER GEOMETRY VALUES
C
  860 WRITE(*,870)
  870 FORMAT('0','Conductivity is defined for a particular'/1X,
     *'prism of the aquifer or aquifer layer and for a part-'/1X,
```

```
   *'icular direction. Therefore, conductivity may change from'/1X,
   *'node to node. Some default values may need to be'/1X,
   *'changed. After each of the following aquifer or aquifer'/1X,
   *'layer hydraulic characteristics and geometries are'/1X,
   *'displayed, enter a real number nodal value.')
880 WRITE(*,890)
890 FORMAT('0','Enter integer number of grid blocks'/1X,
   *'containing nodes whose default values are to be'/1X,
   *'changed  '\)
    READ(*,*)NGB
    DO 1020 L=1,NGB
    WRITE(*,900)L
900 FORMAT('0','Grid block number=',I3)
    WRITE(*,910)
910 FORMAT(1X,'Enter integer grid block left- and right-'/1X,
   *'most column I coordinates separated by a comma  '\)
    READ(*,*)IVG1,IVG2
    WRITE(*,920)
920 FORMAT(1X,'Enter integer grid block upper- and lower-'/1X,
   *'most row J coordinates separated by a comma  '\)
    READ(*,*)JVG1,JVG2
    DO 1010 J=JVG1,JVG2
    DO 1000 I=IVG1,IVG2
    WRITE(*,930)I
930 FORMAT(1X,'I-coordinate of node=',I3)
    WRITE(*,940)J
940 FORMAT(1X,'J-coordinate of node=',I3)
    WRITE(*,950)
950 FORMAT(1X,'Aquifer or aquifer layer horizontal conductivity'/1X,
   *'in Y-direction, vector volume is unit prism centered'/1X,
   *'along I-grid line between I,J and I,J+1, in gpd'/1X,
   *'/sq ft  '\)
    READ(*,*)P1(I,J)
    WRITE(*,960)
960 FORMAT(1X,'Aquifer or aquifer layer horizontal conductivity'/1X,
   *'in X-direction, vector volume is unit prism centered'/1X,
   *'along J-grid line between I,J and I+1,J, in gpd'/1X,
   *'sq ft  '\)
    READ(*,*)P2(I,J)
    WRITE(*,970)
970 FORMAT(1X,'Top of aquifer or aquifer layer '/1X,
   *'elevation in ft  '\)
    READ(*,*)CH(I,J)
    WRITE(*,980)
980 FORMAT(1X,'Base of aquifer or aquifer layer '/1X,
   *'elevation in ft  '\)
    READ(*,*)BOT(I,J)
1000 CONTINUE
1010 CONTINUE
1020 CONTINUE
1030 WRITE(*,1032)
1032 FORMAT('0','Enter 1 to revise default value'/1X,
```

```
     *'changes or 2 to continue data base specification  '\)
      READ(*,*)NRC
      IF(NRC.NE.1.AND.NRC.NE.2) GOTO 1030
      IF(NRC.EQ.1) GOTO 860
 1040 CALL CONSTR(LY1,NN8)
C
C     ADJUST NODAL VALUES FOR VARIABLE GRID
C     BY AVERAGING VARIABLE GRID SPACINGS
C     AND AJUST SQUARE GRID HYDRAULIC
C     CHARACTERISTICS AND FACTORS WITH
C     VECTOR VOLUME TECHNIQUES
C
 1490 DO 1494 J=1,NR
      DO 1492 I=1,NC
      P1(I,J)=P1(I,J)*2*DELX(I)/(DELY(J)+DELY(J+1))
      P2(I,J)=P2(I,J)*2*DELY(J)/(DELX(I)+DELX(I+1))
 1492 CONTINUE
 1494 CONTINUE
C
C     EXPORT DATA BASE
C
 1800 WRITE(*,1810)
 1810 FORMAT('0','Enter 1 to store data base on diskette'/1X,
     *'as ASCII files to be exported for later use'/1X,
     *'with this program or 2 to proceed with program  '\)
      READ(*,*)NST
      IF(NST.NE.1.AND.NST.NE.2)GOTO 1800
      IF(NST.EQ.2)GOTO 1860
      CALL FILEO1(NN8)
C
C     HARD COPY OF NODAL DATA BASE
C
 1860 WRITE(*,1870)
 1870 FORMAT('0','Enter 1 for a hard copy of aquifer or aquifer'/1X,
     *'layer hydraulic characteristics, any pre-existing contam-'/1X,
     *'inant concentration, and any vertical contaminant leakage'/1X,
     *'or 2 for no hard copy  '\)
      READ(*,*)NHC
      IF(NHC.NE.1.AND.NHC.NE.2)GOTO 1860
      IF(NHC.EQ.2)GOTO 2261
      MCX=1
      WRITE(4,1880)
 1880 FORMAT('0','Hydraulic Conductivity in Y-Direction'/1X,
     *'in gpd/sq ft:')
      CALL DATPRN(P1,MCX)
      WRITE(4,1890)
 1890 FORMAT('0','Hydraulic Conductivity in X-Direction'/1X,
     *'in gpd/sq ft:')
      CALL DATPRN(P2,MCX)
      WRITE(4,1900)
 1900 FORMAT('0','Top of aquifer elevation in ft:')
      CALL DATPRN(CH,MCX)
```

```
      WRITE(4,1910)
 1910 FORMAT('0','Base of aquifer elevation in ft:')
      CALL DATPRN(BOT,MCX)
      IF(NPE.EQ.1)THEN
      WRITE(4,1914)
 1914 FORMAT('0','Pre-existing concentration in mg/l:')
      MCX=4
      CALL DATPRN(CONSOR,MCX)
      ENDIF
      IF(NAL.EQ.1)THEN
      WRITE(4,1916)
 1916 FORMAT('0','Simulation factor in gpd/ft:')
      MCX=2
      CALL DATPRN(R,MCX)
      WRITE(4,1918)
 1918 FORMAT('0','Contaminant concentration in sourcebed'/1X,
     *'in mg/l:')
      MCX=4
      CALL DATPRN(CSCC,MCX)
      WRITE(1920)
 1920 FORMAT('0','Elevation of head in sourcebed in ft:')
      MCX=1
      CALL DATPRN(RH,MCX)
      ENDIF
      IF(NLSK.EQ.1.AND.NTT.EQ.2)THEN
      MCX=3
      WRITE(4,1922)
 1922 FORMAT('0','Vertical hydraulic conductivity of aquitard'/1X,
     *'in gpd/sq ft:')
      CALL DATPRN(P3,MCX)
      MCX=1
      WRITE(4,1923)
 1923 FORMAT('0','Thickness of aquitard in ft:')
      CALL DATPRN(THICK3,MCX)
      ENDIF
      IF(NLSK.EQ.1.AND.NTT.EQ.1)THEN
      WRITE(4,1924)
 1924 FORMAT('0','Vertical hydraulic conductivity of aquifer'/1X,
     *'layer 1 in gpd/sq ft:')
      CALL DATPRN(P4,MCX)
      WRITE(4,1925)
 1925 FORMAT('0','Vertical hydraulic conductivity of aquifer'/1X,
     *'layer 2 in gpd/sq ft:')
      CALL DATPRN(P5,MCX)
      WRITE(4,1926)
 1926 FORMAT('0','Aquifer layer 1 thickness in ft:')
      CALL DATPRN(THICK4,MCX)
      WRITE(4,1927)
 1927 FORMAT('0','Aquifer layer 2 thickness in ft:')
      CALL DATPRN(THICK5,MCX)
      ENDIF
C
```

```
C     FILL ARRAYS WITH HEAD ELEVATION VALUES
C
 2261 WRITE(*,2262)
 2262 FORMAT('0','If areal leakage sinks are simulated'/1X,
     *'head data base should reflect average conditions'/1X,
     *'during a time increment following the average travel'/1X,
     *'time through aquitard or between aquifer layers')
      WRITE(*,2264)
 2264 FORMAT('0','Head data base should reflect water level'/1X,
     *'impacts due to any simulated contaminant source or'/1X,
     *'sink.')
 2268 WRITE(*,2270)
 2270 FORMAT('0','Enter 1 for keyboard aquifer head data base'/1X,
     *'input or 2 to import ASCII data base file previously'/1X,
     *'created with program  '\)
      READ(*,*)KXI
      IF(KXI.NE.1.AND.KXI.NE.2)GOTO 2268
      IF(KXI.EQ.1)THEN
 2272 WRITE(*,2274)
 2274 FORMAT('0','Enter 1 to create a head data base with'/1X,
     *'uniform gradient in the X-direction or 2 to continue'/1X,
     *'program. Do not enter 1 unless you previously specified'/1X,
     *'uniform grid spacing and hydraulic conductivity  '\)
      READ(*,*)MHD
      IF(MHD.NE.1.AND.MHD.NE.2)GOTO 2272
      IF(MHD.EQ.1)CALL HEAD(H,GS,PP)
      GOTO 2400
      ENDIF
      IF(KXI.EQ.2)THEN
      CALL FILEI2
      GOTO 2400
      ENDIF
 2300 DO 2328 J=1,NR
      DO 2326 I=1,NC
      WRITE(*,2320)I
 2320 FORMAT('0','I-coordinate of node=',I3)
      WRITE(*,2322)J
 2322 FORMAT(1X,'J-coordinate of node=',I3)
      WRITE(*,2324)
 2324 FORMAT(1X,'Enter real number aquifer or aquifer layer'/1X,
     *'head elevation at node in ft  '\)
      READ(*,*)H(I,J)
 2326 CONTINUE
 2328 CONTINUE
 2332 WRITE(*,2334)
 2334 FORMAT('0','Enter 1 to create an ASCII file of'/1X,
     *'aquifer or aquifer layer head data or 2 for no file  '\)
      READ(*,*)NFL
      IF(NFL.NE.1.AND.NFL.NE.2)GOTO 2332
      IF(NFL.EQ.1)THEN
      CALL FILEO2
      ENDIF
```

```
C
C     CALCULATE INTERNODAL VELOCITIES
C
 2400 DO 2420 J=1,NR
      DO 2410 I=1,NC
      IF(J.NE.NR)V(I,J,1)=-P1(I,J)*(H(I,J+1)-
     *H(I,J))/(DELX(I)*EPOR*7.48052)
      IF(I.NE.NC)V(I,J,2)=-P2(I,J)*(H(I+1,J)-
     *H(I,J))/(DELY(J)*EPOR*7.48052)
 2410 CONTINUE
 2420 CONTINUE
      CALL DISPER
      DISPL=DISPL/RD1
      DISPT=DISPT/RD1
      DO 2424 J=1,NR
      DO 2422 I=1,NC
      CSLUG(I,J)=0.0
 2422 CONTINUE
 2424 CONTINUE
 2425 WRITE(*,2426)
 2426 FORMAT('0','Enter 1 to simulate contamination'/1X,
     *'or 2 to simulate decontamination'/1X,
     *'assuming capture wells are within plume  '\)
      READ(*,*)NSOD
      IF(NSOD.NE.1.AND.NSOD.NE.2)GOTO 2425
C
C     CALL CONTAMINATION SUBROUTINES
C
 2430 CALL CLEAR
      MFL=1
      IF(NN1.EQ.1.OR.NN1.EQ.3)CALL SLUGIN(CSLUG)
      IF(NPE.EQ.1)CALL PRESOR
      IF(NN1.EQ.1.OR.NN1.EQ.3)THEN
      WRITE(*,2432)
 2432 FORMAT('0','COMPUTATIONS ARE IN PROGRESS')
      MFL=2
      ENDIF
      IF(NN1.EQ.3)MFL=1
      CALL ADVAN(DELTA)
      IF(NN1.NE.1.AND.NAL.EQ.1)THEN
      IF(MFL.EQ.1)WRITE(*,2433)
 2433 FORMAT('0','COMPUTATIONS ARE IN PROGRESS')
      MFL=2
      CALL ARESOR
      ENDIF
      IF(NI1.EQ.1)CALL CONTIN(MFL)
      IF(IWS.EQ.1)CALL INJIN(MFL)
      IF(NLC.EQ.1.OR.NN8.EQ.2)CALL LOCSOR(MFL,NN8)
      CALL ARESNK(CSLUG,MCX,NSOD)
      CALL AREA3D(MFL)
      IF(NRAD.EQ.1)PM=PM*(.5)**(TMAP/HL)
      IF(NLSK.EQ.1)GOTO 2610
```

```
      DO 2450 J=1,NR
      DO 2440 I=1,NC
      CONC(I,J)=NPART(I,J)*PM*1E6/(APOR*
     *(CH(I,J)-BOT(I,J))*DELX(I)*DELY(J)*62.4)
      IF(NSOD.EQ.1)CONC(I,J)=CONC(I,J)/RD1
      IF(NSOD.EQ.2)CONC(I,J)=CONC(I,J)*RD1
 2440 CONTINUE
 2450 CONTINUE
C
C     PRINT NODAL CONCENTRATION COMPUTATION RESULTS ON
C     SCREEN
C
 2610 WRITE(*,2620)
 2620 FORMAT('0','Enter 1 for screen display of concentration'/1X,
     *'computation nodal results or 2 for no display  '\)
      READ(*,*)NSCR
      IF(NSCR.NE.1.AND.NSCR.NE.2)GOTO 2610
      IF(NSCR.EQ.2)GOTO 2810
      WRITE(*,2630)
 2630 FORMAT('0','Use Ctrl-Num Lock keys to'/1X,
     *'control screen scrolling')
      PAUSE
      N1=1
      N2=8
      IF(NC.LT.N2)N2=NC
      WRITE(*,2640)TMAP
 2640 FORMAT('0','TIME (DAYS)=',F9.3)
      WRITE(*,2650)
 2650 FORMAT('0','CONCENTRATION AT NODES IN MG/L :')
 2652 WRITE(*,2660)
 2660 FORMAT('0','ROW                          COLUMN'/)
 2670 WRITE(*,2680)(I,I=N1,N2)
 2680 FORMAT(3X,8I9)
      DO 2700 J=1,NR
      WRITE(*,2690)J,(CONC(I,J),I=N1,N2)
 2690 FORMAT(I3,8F9.3)
 2700 CONTINUE
      IF(N2.EQ.NC)GOTO 2810
      N1=N1+8
      N2=N2+8
      IF(NC.LT.N2)N2=NC
      GOTO 2652
C
C     PRINT NODAL PARTICLE DISTRIBUTION ON SCREEN
C
 2810 WRITE(*,2812)
 2812 FORMAT('0','Enter 1 for screen display of particle'/1X,
     *'distribution computation nodal results or 2 for no display  '\)
      READ(*,*)NSCR
      IF(NSCR.NE.1.AND.NSCR.NE.2)GOTO 2810
      IF(NSCR.EQ.2)GOTO 2830
      WRITE(*,2814)
```

```
 2814 FORMAT('0','Use Ctrl-Num Lock keys to'/1X,
     *'control screen scrolling')
      PAUSE
      N1=1
      N2=15
      IF(NC.LT.N2)N2=NC
      WRITE(*,2816)TMAP
 2816 FORMAT('0','TIME (DAYS)=',F12.2)
      WRITE(*,2818)
 2818 FORMAT('0','PARTICLES AT NODES:')
 2819 WRITE(*,2820)
 2820 FORMAT('0','ROW                              COLUMN'/)
 2822 WRITE(*,2824)(I,I=N1,N2)
 2824 FORMAT(3X,15I5)
      DO 2828 J=1,NR
      WRITE(*,2826)J,(NPART(I,J),I=N1,N2)
 2826 FORMAT(I3,15I5)
 2828 CONTINUE
      IF(N2.EQ.NC)GOTO 2830
      N1=N1+15
      N2=N2+15
      IF(NC.LT.N2)N2=NC
      GOTO 2819
C
C     CREATE A HARD COPY OF NODAL CONCENTRATION
C     COMPUTATION RESULTS
C
 2830 WRITE(*,2850)
 2850 FORMAT('0','Enter 1 for a hard copy of nodal concentration'/1X,
     *'computation results and sink concentration computation'/1X,
     *'results or 2 to continue program  '\)
      READ(*,*)MCS
      IF(MCS.NE.1.AND.MCS.NE.2)GOTO 2830
      IF(MCS.EQ.1)THEN
      CALL CONMAP(NSOD)
      CALL SUMMRY
      CALL SNKCON(NSOD)
      ENDIF
 2852 WRITE(*,2854)
 2854 FORMAT('0','Enter 1 for hard copy of particle location'/1X,
     *'map or 2 for no display  '\)
      READ(*,*)NDM
      IF(NDM.NE.1.AND.NDM.NE.2)GOTO 2852
      IF(NDM.EQ.1)CALL MAP
C
C     EXPORT NODAL CONCENTRATION COMPUTATION RESULTS
C
 2910 WRITE(*,2920)
 2920 FORMAT('0','Enter 1 to create an ASCII file of'/1X,
     *'concentration computation nodal results by rows with'/1X,
     *'one concentration on a line or 2 for no file  '\)
      READ(*,*)NFL
```

```
      IF(NFL.NE.1.AND.NFL.NE.2)GOTO 2910
      IF(NFL.EQ.2)GOTO 2990
      WRITE(*,2930)
2930 FORMAT('0','Enter a valid computation nodal result'/1X,
     *'filename with extension DAT such as B:FILE2.DAT'/1X,
     *'in apostrophes  '\)
      READ(*,*)FNAME
      OPEN(6,FILE=FNAME,STATUS='NEW')
      DO 2970 J=1,NR
      DO 2960 I=1,NC
      WRITE(6,2940)CONC(I,J)
2940 FORMAT(1X,F9.3)
2960 CONTINUE
2970 CONTINUE
2980 CLOSE (6)
2990 CALL XYZDTC(CONC)
      CALL PROFIL
      CLOSE (4)
3000 END
```

```
C
      SUBROUTINE ADD(XX,YY)
C
C     THIS SUBROUTINE IS CALLED WHEN PARTICLES ARE
C     TO BE ADDED TO THE SYSTEM. IF THE NUMBER OF
C     PARTICLES TO BE ADDED IS GREATER THAN MAXP
C     THEN PM CHANGES WHILE NP STAYS THE SAME
C
C     METACOMMANDS
C
$LARGE
C
C     DECLARATIONS
C
      COMMON/NSUB7/NP,NPART(50,50)
      COMMON/NSUB6/MAXP,NSORS,NIJ,NSSN,MARK(50,50),NRAD
      COMMON/AREA7/PM,DISPL,DISPT,EPOR,APOR
      COMMON/AREA8/TMAP,ANC,ANR,TABLE(124)
      COMMON/AREA5/X(5001),Y(5001),V(50,50,2)
C
      NP=NP+1
      X(NP)=XX
      Y(NP)=YY
      IF(NP.LT.MAXP)RETURN
      K=RAN(0)*MAXP+1.0
      X(K)=X(NP)
      Y(K)=Y(NP)
      PM=PM*MAXP/(MAXP-1)
      NP=NP-1
      END
```

```
C
      SUBROUTINE ADVAN(DELTA)
C
C     THIS SUBROUTINE ADVANCES SLUG SOURCE PARTICLES TO
C     THEIR NEW POSITIONS DURING TIME DELTA
C
C     METACOMMANDS
C
$LARGE
C
C     DECLARATIONS
C
      COMMON/NSUB7/NP,NPART(50,50)
      COMMON/NSUB6/MAXP,NSORS,NIJ,NSSN,MARK(50,50),NRAD
      COMMON/AREA7/PM,DISPL,DISPT,EPOR,APOR
      COMMON/AREA8/TMAP,ANC,ANR,TABLE(124)
      COMMON/AREA5/X(5001),Y(5001),V(50,50,2)
      COMMON/AREA3/RD1,HL,PLEAK(50),Q(50,50),RD1V
      COMMON/NSUB1/NC,NR
      COMMON/AREA2/DELX(51),DELY(51),P1(50,50),P2(50,50)
C
      TMAP=TMAP+DELTA
      K=0
    1 K=K+1
    2 IF(K.GT.NP)RETURN
      IF(MOVE(X(K),Y(K),DELTA).EQ.0)GOTO 1
      X(K)=X(NP)
      Y(K)=Y(NP)
      NP=NP-1
      GO TO 2
      END
```

```
C
      FUNCTION ANORM(K)
C
C      THIS FUNCTION PRODUCES A RANDOM NUMBER
C      (-6,+6) NORMAL DISTRIBUTION
C
      ANORM=-6.0
      DO 10 I=1,12
      ANORM=ANORM+RAN(0)
   10 CONTINUE
      END
```

```
C
      SUBROUTINE AREA3D(MFL)
C
C     THIS SUBROUTINE DEFINES CONTAMINANT LOAD FROM AREAL
C     LEAKAGE SINKS AND DISTRIBUTES PARTICLES RANDOMLY IN
C     SPACE AND TIME
C
C     METACOMMANDS
C
$LARGE
C
C     DECLARATIONS
C
      CHARACTER*14FNAME
      COMMON/NSUB1/NC,NR
      COMMON/AREA6/DELTA,THICKC(50,50),PS(50,50)
      COMMON/NSUB9/LEK,NRE,NCS
      COMMON/NSUB7/NP,NPT(50,50)
      COMMON/NSUB6/MAXP,NSORS,NIJ,NSSN,MARK(50,50),NRAD
      COMMON/AREA3/RD1,HL,PLEAK(50),Q(50,50),RD1V
      COMMON/AREA7/PM,DISPL,DISPT,EPOR,APOR
      COMMON/AREA8/TMAP,ANC,ANR,TABLE(124)
      COMMON/AREA2/DELX(51),DELY(51),P1(50,50),P2(50,50)
      COMMON/AREA5/X(5001),Y(5001),V(50,50,2)
C
      IF(LEK.EQ.1)THEN
 1111 WRITE(*,1112)
 1112 FORMAT('0','Enter 1 for keyboard areal contaminant'/1X,
     *'source nodal load input or 2 import ASCII data base'/1X,
     *'file previously created with this program  '\)
      READ(*,*)MX
      IF(MX.NE.1.AND.MX.NE.2)GOTO 1111
      IF(MX.EQ.2)THEN
      WRITE(*,1113)
 1113 FORMAT('0','Enter a valid areal contaminant source'/1X,
     *'load filename with extension DAT such as B:FILE6.DAT'/1X,
     *'in apostrophes  '\)
      READ(*,*)FNAME
      OPEN(6,FILE=FNAME)
      DO 1116 J=1,NR
      DO 1115 I=1,NC
      READ(6,1114)PS(I,J)
 1114 FORMAT(1X,F9.3)
 1115 CONTINUE
 1116 CONTINUE
      CLOSE (6)
```

```
      ENDIF
      IF(MX.EQ.1)THEN
 1117 WRITE(*,1118)
 1118 FORMAT('0','Enter integer number of grid blocks'/1X,
     *'containing nodes whose aquifer or aquifer layer'/1X,
     *'areal contaminant source load values are to be'/1X,
     *'specified  '\)
      READ(*,*)NGB
      DO 1127 L=1,NGB
      WRITE(*,1119)L
 1119 FORMAT('0','Grid block number=',I3)
      WRITE(*,1120)
 1120 FORMAT(1X,'Enter integer grid block left- and right-'/1X,
     *'most column I coordinates separated by a comma  '\)
      READ(*,*)IVG1,IVG2
      WRITE(*,1121)
 1121 FORMAT(1X,'Enter integer grid block upper- and lower-'/1X,
     *'most row J coodinates separated by a comma  '\)
      READ(*,*)JVG1,JVG2
      DO 1126 J=JVG1,JVG2
      DO 1125 I=IVG1,IVG2
      WRITE(*,1122)I
 1122 FORMAT('0','I-coordinate of node=',I3)
      WRITE(*,1123)J
 1123 FORMAT(1X,'J-coordinate of node=',I3)
      WRITE(*,1124)
 1124 FORMAT(1X,'Enter real number aquifer or aquifer'/1X,
     *'layer areal contaminant source nodal load'/1X,
     *'in pounds  '\)
      READ(*,*)PS(I,J)
 1125 CONTINUE
 1126 CONTINUE
 1127 CONTINUE
      ENDIF
      IF(MFL.EQ.1)WRITE(*,1128)
 1128 FORMAT('0','COMPUTATIONS ARE IN PROGRESS')
      MFL=2
      DO 1131 J=1,NR
      DO 1130 I=1,NC
      IF(PS(I,J).EQ.0.0)GOTO 1130
      P=PS(I,J)
 1129 IF(RAN(0).GT.P/PM)GOTO 1130
      P=P-PM
      XX=((I-1)+.5)+1.0*RAN(0)
      YY=((J-1)+.5)+1.0*RAN(0)
      DE=DELTA*RAN(0)
      IF(MOVE(XX,YY,DE).EQ.0)CALL ADD(XX,YY)
      GOTO 1129
 1130 CONTINUE
 1131 CONTINUE
      ENDIF
      END
```

```
C
      SUBROUTINE ARESNK(CSLUG,MCX,NSOD)
C
C     THIS SUBROUTINE DEFINES CONTAMINANT CONCENTRATIONS
C     WITH AREAL LEAKAGE SINKS
C
C     METACOMMANDS
C
$LARGE
C
C     DECLARATIONS
C
      CHARACTER*14FNAME
      DIMENSION CSINK(50,50),CSLUG(50,50)
      COMMON/NSUB1/NC,NR
      COMMON/NSUB8/NAL,NLSK,NTT,NPE,NN1,IWS,NLC,NI1
      COMMON/ASUB1/CONSNK(50,50),CONC(50,50),
     *CSCC(50,50),CONSOR(50,50)
      COMMON/ASUB2/CH(50,50),BOT(50,50),H(50,50)
      COMMON/AREA2/DELX(51),DELY(51),P1(50,50),P2(50,50)
      COMMON/AREA3/RD1,HL,PLEAK(50),Q(50,50),RD1V
      COMMON/AREA6/DELTA,THICKC(50,50),PS(50,50)
      COMMON/AREA7/PM,DISPL,DISPT,EPOR,APOR
      COMMON/ASUB3/P3(50,50),P4(50,50),P5(50,50),THICK3(50,50)
      COMMON/ASUB4/THICK4(50,50),THICK5(50,50),H1(50,50),H2(50,50)
      COMMON/NSUB7/NP,NPART(50,50)
      COMMON/NSUB6/MAXP,NSORS,NIJ,NSSN,MARK(50,50),NRAD
      OPEN(5,FILE='LPT1:')
      IF(NLSK.EQ.1)THEN
      DELT=0.0
      DO 2410 J=1,NR
      DO 2408 I=1,NC
      CONC(I,J)=NPART(I,J)*PM*1E6/(APOR*(CH(I,J)-BOT(I,J))*
     *DELX(I)*DELY(J)*62.4)
      IF(NSOD.EQ.1)CONC(I,J)=CONC(I,J)/RD1
      IF(NSOD.EQ.2)CONC(I,J)=CONC(I,J)*RD1
      CSINK(I,J)=CONSOR(I,J)+CSLUG(I,J)
 2408 CONTINUE
 2410 CONTINUE
      IF(NTT.EQ.2)THEN
      WRITE(*,2411)
 2411 FORMAT('0','Enter real number effective porosity of'/1X,
     *'aquitard as a decimal, usually ranges from .04 to'/1X,
     *'.4  '\)
      READ(*,*)EPORA
      DO 2422 J=1,NR
```

```
      DO 2420 I=1,NC
      CONC1=CONC(I,J)
      RATE2=ABS(H1(I,J)-H(I,J))*(P3(I,J)/THICK3(I,J))
      F1=RATE2*DELTA*8.3453/(APOR*ABS(CH(I,J)-BOT(I,J))*62.4)
      C1=2*CONC(I,J)-CSINK(I,J)*F1
      IF(C1.LE.0.0)THEN
      CONC(I,J)=0.0
      GOTO 2412
      ENDIF
      CONC(I,J)=C1/(F1+2)
2412  PS(I,J)=(CONC1-CONC(I,J))*APOR*(CH(I,J)-BOT(I,J))*DELX(I)*
     *DELY(J)*62.4/1E6
      PS(I,J)=PS(I,J)/RD1V
      DELT=DELT+7.48*EPORA*RD1V*THICK3(I,J)**2/(P3(I,J)*ABS
     *(H1(I,J)-H(I,J)))
2420  CONTINUE
2422  CONTINUE
      GOTO 2460
      ENDIF
      DO 2458 J=1,NR
      DO 2456 I=1,NC
      CONC1=CONC(I,J)
      MAV=(THICK4(I,J)+THICK5(I,J))/2
      PAV=(THICK4(I,J)+THICK5(I,J))/((THICK4(I,J)/P4(I,J))+
     *(THICK5(I,J)/P5(I,J)))
      RATE2=ABS(H2(I,J)-H(I,J))*16.6906*DELTA/(THICK4(I,J)/
     *P4(I,J)+THICK5(I,J)/P5(I,J))
      F1=RATE2/(APOR*THICK4(I,J)*62.4)
      C1=2*CONC(I,J)-CSINK(I,J)*F1
      IF(C1.LE.0.0)THEN
      CONC(I,J)=0.0
      GOTO 2424
      ENDIF
      CONC(I,J)=C1/(F1+2)
2424  PS(I,J)=(CONC1-CONC(I,J))*APOR*THICK4(I,J)*DELX(I)*
     *DELY(J)*62.4/1E6
      PS(I,J)=PS(I,J)/RD1
      DELT=DELT+7.48*EPOR*RD1*MAV**2/(PAV*ABS(H2(I,J)-H(I,J)))
2456  CONTINUE
2458  CONTINUE
2460  DELTAS=DELT/(NC*NR)
      DO 2462 J=1,NR
      DO 2461 I=1,NC
      IF(NRAD.EQ.1)PS(I,J)=PS(I,J)*(.5)**(DELTAS/HL)
2461  CONTINUE
2462  CONTINUE
      WRITE(*,2463)DELTAS
2463  FORMAT('0','Average contaminant travel time'/1X,
     *'through aquitard or between aquifer layers'/1X,
     *'assuming steady state vertical head loss and'/1X,
     *'vertical dispersion is negligible in days=',/1X,
     *F9.2)
```

```
      WRITE(*,2464)
2464 FORMAT(1X,'Heads in aquifer or aquifer layer'/1X,
    *'receiving contaminant load from areal leakage'/1X,
    *'sinks should reflect average head conditions'/1X,
    *'during a time increment after the above travel'/1X,
    *'time. The time increment should be the same as'/1X,
    *'that used in the areal leakage sink data base')
      PAUSE
2466 WRITE(*,2468)
2468 FORMAT('0','Enter 1 to create an ASCII file of'/1X,
    *'areal leakage sink contaminant nodal load by rows'/1X,
    *'with one load on a line or 2 for no file  '\)
      READ(*,*)NLO
      IF(NLO.NE.1.AND.NLO.NE.2)GOTO 2466
      IF(NLO.EQ.1)THEN
      WRITE(*,2470)
2470 FORMAT('0','Enter a valid nodal contaminant load'/1X,
    *'filename with extension DAT such as B:FILE3.DAT'/1X,
    *'in apostrophes  '\)
      READ(*,*)FNAME
      OPEN(6,FILE=FNAME,STATUS='NEW')
      DO 2476 J=1,NR
      DO 2474 I=1,NC
      WRITE(6,2472)PS(I,J)
2472 FORMAT(1X,F9.3)
2474 CONTINUE
2476 CONTINUE
      CLOSE (6)
      ENDIF
2478 WRITE(*,2480)
2480 FORMAT('0','Enter 1 for hard copy of areal leakage'/1X,
    *'sink contaminant nodal load or 2 for no hard copy  '\)
      READ(*,*)MH5
      IF(MH5.NE.1.AND.MH5.NE.2)GOTO 2478
      IF(MH5.EQ.1)THEN
      WRITE(5,2482)
2482 FORMAT('0','Areal leakage sink contaminant nodal'/1X,
    *'load in pounds:')
      MCX=4
      CALL DATPRN(PS,MCX)
      ENDIF
      ENDIF
      CLOSE (5)
      END
```

```
C
      SUBROUTINE ARESOR
C
C     THIS SUBROUTINE GENERATES PARTICLES TO SIMULATE
C     AREAL CONTAMINANT LEAKAGE THROUGH AQUIFER LAYERS OR
C     AQUITARDS. NODAL LEAKAGE AND CONTAMINANT LOAD ARE
C     DEFINED AND PARTICLES ARE RANDOMLY DISTRIBUTED IN
C     AREA CENTERED AT NODES AND IN TIME
C
C     METACOMMANDS
C
$LARGE
C
C     DECLARATIONS
C
      COMMON/NSUB1/NC,NR
      COMMON/AREA6/DELTA,THICKC(50,50),PS(50,50)
      COMMON/ASUB1/CONSNK(50,50),CONC(50,50),
     *CSCC(50,50),CONSOR(50,50)
      COMMON/AREA1/RH(50,50),R(50,50),RD(50,50),RS(50,50)
      COMMON/NSUB7/NP,NPART(50,50)
      COMMON/NSUB6/MAXP,NSORS,NIJ,NSSN,MARK(50,50),NRAD
      COMMON/AREA7/PM,DISPL,DISPT,EPOR,APOR
      COMMON/ASUB2/CH(50,50),BOT(50,50),H(50,50)
      COMMON/AREA5/X(5001),Y(5001),V(50,50,2)
      COMMON/AREA8/TMAP,ANC,ANR,TABLE(124)
      COMMON/AREA3/RD1,HL,PLEAK(50),Q(50,50),RD1V
      COMMON/AREA2/DELX(51),DELY(51),P1(50,50),P2(50,50)
C
      DO 120 J=1,NR
      DO 110 I=1,NC
      RATE=(RH(I,J)-H(I,J))*R(I,J)
      PL=RATE*DELTA*CSCC(I,J)*8.3453/1E6
      P=PL
  100 IF(RAN(0).GT.P/PM)GOTO 110
      P=P-PM
      XX=((I-1)+.5)+1.0*RAN(0)
      YY=((J-1)+.5)+1.0*RAN(0)
      DE=DELTA*RAN(0)
      IF(MOVE(XX,YY,DE).EQ.0)CALL ADD(XX,YY)
      GOTO 100
  110 CONTINUE
  120 CONTINUE
  130 END
```

```
C
      SUBROUTINE CLEAR
C
C     THIS SUBROUTINE CLEARS OUT NPART AND TABLE ARRAYS
C
C     METACOMMANDS
C
$DEBUG
C
C     DECLARATIONS
C
      COMMON/NSUB1/NC,NR
      COMMON/NSUB7/NP,NPART(50,50)
      COMMON/AREA8/TMAP,ANC,ANR,TABLE(124)
C
      DO 20 J=1,NR
      DO 10 I=1,NC
      NPART(I,J)=0
   10 CONTINUE
   20 CONTINUE
      DO 30 I=1,124
      TABLE(I)=0.0
   30 CONTINUE
      END
```

```
C
      SUBROUTINE CONMAP(NSOD)
C
C     THIS SUBROUTINE CONVERTS NPART TO
C     CONCENTRATION AND PROVIDES A HARD COPY
C     OF NODAL CONCENTRATION
C
C     METACOMMANDS
C
$LARGE
C
C     DECLARATIONS
C
      COMMON/NSUB1/NC,NR
      COMMON/NSUB7/NP,NPART(50,50)
      COMMON/NSUB8/NAL,NLSK,NTT,NPE,NN1,IWS,NLC,NI1
      COMMON/AREA1/RH(50,50),R(50,50),RD(50,50),RS(50,50)
      COMMON/AREA3/RD1,HL,PLEAK(50),Q(50,50),RD1V
      COMMON/AREA8/TMAP,ANC,ANR,TABLE(124)
      COMMON/AREA6/DELTA,THICKC(50,50),PS(50,50)
      COMMON/AREA7/PM,DISPL,DISPT,EPOR,APOR
      COMMON/ASUB2/CH(50,50),BOT(50,50),H(50,50)
      COMMON/AREA2/DELX(51),DELY(51),P1(50,50),P2(50,50)
      COMMON/ASUB1/CONSNK(50,50),CONC(50,50),
     *CSCC(50,50),CONSOR(50,50)
C
      OPEN(5,FILE='LPT1:')
      IF(NLSK.EQ.1)GOTO 42
      DO 20 J=1,NR
      DO 10 I=1,NC
      CONC(I,J)=0.0
   10 CONTINUE
   20 CONTINUE
      DO 40 J=1,NR
      DO 30 I=1,NC
      CONC(I,J)=NPART(I,J)*PM*1E6/(APOR*
     *(CH(I,J)-BOT(I,J))*DELX(I)*DELY(J)*62.4)
      IF(NSOD.EQ.1)CONC(I,J)=CONC(I,J)/RD1
      IF(NSOD.EQ.2)CONC(I,J)=CONC(I,J)*RD1
   30 CONTINUE
   40 CONTINUE
   42 N1=1
      N2=8
      IF(NC.LT.N2)N2=NC
   48 WRITE(5,50)TMAP
   50 FORMAT('0','TIME (DAYS)=',F9.3)
```

```
      WRITE(5,60)
   60 FORMAT('0','CONCENTRATION VALUES (MG/L):')
   62 WRITE(5,70)
   70 FORMAT('0','ROW                              COLUMN')
   72 WRITE(5,74)(I,I=N1,N2)
   74 FORMAT(3X,8I9)
      DO 100 J=1,NR
      WRITE(5,76)J,(CONC(I,J),I=N1,N2)
   76 FORMAT(I3,8F9.3)
  100 CONTINUE
      IF(N2.EQ.NC)GO TO 200
      N1=N1+8
      N2=N2+8
      IF(NC.LT.N2)N2=NC
      GO TO 62
      CLOSE (5)
  200 END
```

```
C
      SUBROUTINE CONSTR(LY1,NN8)
C
C     THIS SUBROUTINE DEFINES SOURCE AND SINK GEOMETRY
C     AND STRENGTHS AND SORPTION AND RADIOACTIVE FACTORS
C
C     METACOMMANDS
C
$LARGE
C
C     DECLARATIONS
C
      CHARACTER*14FNAME
      COMMON/NSUB1/NC,NR
      COMMON/NSUB2/NPUMP,NPITS
      COMMON/NSUB3/IJJ(24),JIJ(24)
      COMMON/NSUB4/IP(24),JP(24),ISS(100),JSS(100)
      COMMON/NSUB6/MAXP,NSORS,NIJ,NSSN,MARK(50,50),NRAD
      COMMON/NSUB7/NP,NPT(50,50)
      COMMON/NSUB8/NAL,NLSK,NTT,NPE,NN1,IWS,NLC,NI1
      COMMON/NSUB9/LEK,NRE,NCS
      COMMON/ASUB1/CONSNK(50,50),CONC(50,50),
     *CSCC(50,50),CONSOR(50,50)
      COMMON/AREA1/RH(50,50),R(50,50),RD(50,50),RS(50,50)
      COMMON/AREA2/DELX(51),DELY(51),P1(50,50),P2(50,50)
      COMMON/AREA3/RD1,HL,PLEAK(50),Q(50,50),RD1V
      COMMON/AREA5/X(5001),Y(5001),V(50,50,2)
      COMMON/AREA6/DELTA,THICKC(50,50),PS(50,50)
      COMMON/AREA7/PM,DISPL,DISPT,EPOR,APOR
      COMMON/AREA8/TMAP,ANC,ANR,TABLE(124)
      COMMON/ASUB3/P3(50,50),P4(50,50),P5(50,50),THICK3(50,50)
      COMMON/ASUB4/THICK4(50,50),THICK5(50,50),H1(50,50),H2(50,50)
      OPEN(5,FILE='LPT1:')
C
C     DEFINE PRE-EXISTING CONTAMINANT CONDITIONS
C
      IF(NPE.EQ.1)THEN
  800 WRITE(*,810)
  810 FORMAT('0','Enter 1 for keyboard pre-existing contaminant'/1X,
     *'condition data base input or 2 to import ASCII pre-existing'/1X,
     *'contaminant condition data base  '\)
      READ(*,*)KDB
      IF(KDB.NE.1.AND.KDB.NE.2)GOTO 800
      IF(KDB.EQ.1)GOTO 882
      CALL FILEI3
      GOTO 1020
```

```
 882 WRITE(*,884)
 884 FORMAT('0','Enter 1 for uniform pre-existing concentration'/1X,
    *'at each node or 2 for nonuniform pre-existing'/1X,
    *'concentration  '\)
     READ(*,*)NPC
     IF(NPC.NE.1.AND.NPC.NE.2)GOTO 882
     IF (NPC.EQ.2)GOTO 890
     WRITE(*,885)
 885 FORMAT('0','Enter real number uniform pre-existing'/1X,
    *'concentration at all nodes in mg/l  '\)
     READ(*,*)UC
     DO 887 J=1,NR
     DO 886 I=1,NC
     CONSOR(I,J)=UC
 886 CONTINUE
 887 CONTINUE
     GOTO 1010
 890 DO 1008 J=1,NR
     DO 1006 I=1,NC
     WRITE(*,1000)I
1000 FORMAT('0','I-coordinate of node=',I3)
     WRITE(*,1002)J
1002 FORMAT(1X,'J-coordinate of node=',I3)
     WRITE(*,1004)
1004 FORMAT(1X,'Enter real number pre-existing concentration'/1X,
    *'at node in mg/l  '\)
     READ(*,*)CONSOR(I,J)
1006 CONTINUE
1008 CONTINUE
1010 WRITE(*,1012)
1012 FORMAT('0','Enter 1 to export pre-existing contaminant'/1X,
    *'condition data base  '\)
     READ(*,*)MEC
     IF(MEC.NE.1.AND.MEC.NE.2)GOTO 1010
     IF(MEC.EQ.1)CALL FILEO3
1020 ENDIF
     WRITE(*,1025)
1025 FORMAT('0','A fatal error requiring that diskettes be'/1X,
    *'removed from drives, the computer shut off, and the'/1X,
    *'program rerun will occur if contaminant source or sink'/1X,
    *'geometry and/or strength are unrealistic and the contam-'/1X,
    *'inant mass load divided by the particle mass exceeds'/1X,
    *'5000. Please check reasonableness of source and sink'/1X,
    *'data base to prevent a lockup.')
1048 WRITE(*,1049)
1049 FORMAT('0','Enter 1 to simulate slug (injection only'/1X,
    *'at start of time increment) contaminant sources, 2'/1X,
    *'to simulate continuous (random or uniform injection'/1X,
    *'during entire time increment) contaminant sources,'/1X,
    *'or 3 to simulate both slug and continuous sources.  '\)
     READ(*,*)NN1
     IF(NN1.NE.1.AND.NN1.NE.2.AND.NN1.NE.3)GOTO 1048
```

```
      IF(NN1.EQ.2)NSORS=0
      IF(NN1.EQ.1.OR.NN1.EQ.3)THEN
      WRITE(*,1050)
 1050 FORMAT(1X,'Enter integer number of slug sources'/1X,
     *'(must be < 25)  '\)
      READ(*,*)NSORS
      ENDIF
      IF(LY1.EQ.2)GOTO 1053
      WRITE(*,1052)
 1052 FORMAT('0','The total contaminant load (mass at 100'/1X,
     *'percent concentration) is the number of particles'/1X,
     *'simulating contaminant concentration within a grid space'/1X,
     *'aquifer volume multiplied by the particle mass. Choice'/1X,
     *'of the number of particles per grid space depends upon'/1X,
     *'the concentration resolution desired. Favorable'/1X,
     *'resolution is usually obtained with 200 particles per'/1X,
     *'grid space. The particle mass may be estimated with the'/1X,
     *'following equation: Particle Mass (pounds) equals'/1X,
     *'Desired Concentration Resolution (mg/l) x Grid Spacing'/1X,
     *'In X-Direction (ft) x Grid Spacing In J-Direction (ft)'/1X,
     *'x Aquifer Thickness (ft) x 62.4 divided by 2000000.')
 1053 WRITE(*,1054)
 1054 FORMAT('0','Enter real number particle mass in pounds  '\)
      READ(*,*)PM
      WRITE(5,1055)PM
 1055 FORMAT(1X,'Particle mass in pounds=',F10.3)
      MAXP=5000
      WRITE(5,1057)MAXP
 1057 FORMAT(1X,'Maximum number of particles in model=',I5)
C
C     DEFINE SORPTION AND RADIOACTIVE DECAY CONSTANTS
C
      RD1=1.0
 1058 WRITE(*,1060)
 1060 FORMAT('0','Enter 1 to simulate contaminant sorption'/1X,
     *'in aquifer or aquifer layer or 2 for no sorption  '\)
      READ(*,*)NRE
      IF(NRE.NE.1..AND.NRE.NE.2)GOTO 1058
      IF(NRE.EQ.1)THEN
      IF(LY1.EQ.2)GOTO 1063
      WRITE(*,1062)
 1062 FORMAT('0','It is assumed that sorption reactions'/1X,
     *'are fast and reversible and the isotherm is linear.'/1X,
     *'It is further assumed that there is only one'/1X,
     *'contaminant in solution and no interaction'/1X,
     *'of contaminants')
 1063 WRITE(*,1064)
 1064 FORMAT('0','Enter real number distribution coefficient'/1X,
     *'of aquifer or aquifer layer in ml/g, usually ranges'/1X,
     *'from 1E-5 to 1E6 ml/g  '\)
      READ(*,*)DK
      WRITE(5,1066)DK
```

```
 1066 FORMAT(1X,'Distribution coefficient in ml/g=',E12.5)
      WRITE(*,1068)
 1068 FORMAT(1X,'Enter real number bulk mass density of dry'/1X,
     *'aquifer skeleton in g/cu cm, usually ranges from .7'/1X,
     *'to 3.2 g/cu cm  '\)
      READ(*,*)RHO
      WRITE(5,1070)RHO
 1070 FORMAT(1X,'Bulk mass density of aquifer skeleton'/1X,
     *'in g/cu cm=',F5.2)
      RD1=1+(RHO/EPOR)*DK
      WRITE(*,1071)RD1
 1071 FORMAT(1X,'Aquifer or aquifer layer retardation factor=',F9.2)
      WRITE(5,1072)RD1
 1072 FORMAT(1X,'Aquifer or aquifer layer retardation factor=',F9.2)
      ENDIF
 1073 WRITE(*,1074)
 1074 FORMAT('0','Enter 1 to simulate radioactive isotope'/1X,
     *'decay in aquifer or aquifer layer or 2 for no decay  '\)
      READ(*,*)NRAD
      IF(NRAD.NE.1.AND.NRAD.NE.2)GOTO 1073
      HL=0.0
      IF(NRAD.EQ.1)THEN
      WRITE(*,1076)
 1076 FORMAT('0','Enter real number half-life of radioactive'/1X,
     *'isotope in days, usually ranges from a few days to'/1X,
     *'millions of years  '\)
      READ(*,*)HL
      WRITE(5,1078)HL
 1078 FORMAT('0','Half-life of radioactive isotope in'/1X,
     *'days in aquifer or aquifer layer =',F15.2)
      ENDIF
      IF(NN1.EQ.1)GOTO 1161
C
C     DEFINE INJECTION WELL LOCATION
C
 1080 WRITE(*,1082)
 1082 FORMAT('0','Enter 1 to simulate injection well continuous'/1X,
     *'source or 2 for no such source  '\)
      READ(*,*)IWS
      IF(IWS.NE.1.AND.IWS.NE.2)GOTO 1080
      IF(IWS.EQ.2)NIJ=0
      IF(IWS.EQ.1)THEN
      WRITE(*,1084)
 1084 FORMAT('0','Enter integer number of injection wells'/1X,
     *'(must be < 51)  '\)
      READ(*,*)NIJ
      DO 1108 I=1,NIJ
      WRITE(*,1086)I
 1086 FORMAT(1X,'Injection well number=',I3)
      WRITE(5,1088)I
 1088 FORMAT('0','Injection well number=',I3)
      WRITE(*,1100)
```

```
 1100 FORMAT(1X,'Enter integer I-coordinate of injection'/1X,
     *'well  '\)
      READ(*,*)IJJ(I)
      WRITE(5,1102)IJJ(I)
 1102 FORMAT(1X,'I-coordinate of injection well=',I3)
      WRITE(*,1104)
 1104 FORMAT(1X,'Enter integer J-coordinate of injection'/1X,
     *'well  '\)
      READ(*,*)JIJ(I)
      WRITE(5,1106)JIJ(I)
 1106 FORMAT(1X,'J-coordinate of injection well=',I3)
 1108 CONTINUE
      ENDIF
 1109 WRITE(*,1110)
 1110 FORMAT('0','Enter 1 to simulate contaminant source load'/1X,
     *'from areal leakage sinks or 2 for no such source. Areal'/1X,
     *'leakage sinks means areal contaminant migration from'/1X,
     *'one aquifer layer to an adjacent aquifer layer or areal'/1X,
     *'contaminant migration from one aquifer through an adjacent'/1X,
     *'aquitard and into another aquifer  '\)
      READ(*,*)LEK
      IF(LEK.NE.1.AND.LEK.NE.2)GOTO 1109
      IF(LEK.EQ.1)GOTO 1144
C
C     DEFINE AREAL OR LOCAL CONTAMINANT LEAKAGE
C
 1121 WRITE(*,1122)
 1122 FORMAT('0','Enter 1 to simulate areal contaminant source'/1X,
     *'leakage through aquitard assuming sourcebed head, aquitard'/1X,
     *'hydraulic characteristics, and contaminant concentration'/1X,
     *'are variable in space; aquifer head is above aquitard'/1X,
     *'base; and sourcebed contaminant mass is not depleted.'/1X,
     *'If leakage rate and contaminant concentration are low '/1X,
     *'source may be negligible. Enter 2 for no such source'/1X,
     *'or if area sinks are simulated  '\)
      READ(*,*)NAL
      IF(NAL.NE.1.AND.NAL.NE.2)GOTO 1121
      IF(NAL.EQ.1)THEN
 1124 WRITE(*,1125)
 1125 FORMAT('0','Enter 1 for keyboard areal source data base'/1X,
     *'input or 2 to import ASCII data base file previously'/1X,
     *'created with this program  '\)
      READ(*,*)LAR
      IF(LAR.NE.1.AND.LAR.NE.2)GOTO 1124
      IF (LAR.EQ.2)THEN
      WRITE(*,1126)
 1126 FORMAT('0','Enter a valid areal source data base filename'/1X,
     *'with extension DAT such as B:FILE6.DAT in apostrophes  '\)
      READ(*,*)FNAME
      OPEN(6,FILE=FNAME)
      DO 1129 J=1,NR
      DO 1128 I=1,NC
```

```
      READ(6,1127)RH(I,J),R(I,J),CSCC(I,J)
1127  FORMAT(1X,F9.2,F12.2,F9.3)
1128  CONTINUE
1129  CONTINUE
      GOTO 1143
      CLOSE (6)
      ENDIF
      DO 1136 J=1,NR
      DO 1135 I=1,NC
      WRITE(*,1130)I,J
1130  FORMAT('0','Node I,J coordinates=',I3,',',I3)
      WRITE(*,1131)
1131  FORMAT(1X,'Enter real number elevation of sourcebed'/1X,
     *'head in ft  '\)
      READ(*,*)RH(I,J)
      WRITE(*,1132)
1132  FORMAT(1X,'Enter real number aquitard vertical'/1X,
     *'hydraulic conductivity at node in gpd/sq ft, usually'/1X,
     *'ranges from 1E-7 to 10 gpd/sq ft  '\)
      READ(*,*)PA
      WRITE(*,1133)
1133  FORMAT(1X,'Enter real number aquitard thickness'/1X,
     *'at node in ft  '\)
      READ(*,*)AT
      WRITE(*,1134)
1134  FORMAT(1X,'Enter real number sourcebed contaminant'/1X,
     *'concentration at node in mg/l  '\)
      READ(*,*)CSCC(I,J)
      R(I,J)=(PA/AT)*DELX(I)*DELY(J)
1135  CONTINUE
1136  CONTINUE
1137  WRITE(*,1138)
1138  FORMAT('0','Enter 1 to export areal source data base to'/1X,
     *'a file or 2 for no file  '\)
      READ(*,*)LA1
      IF(LA1.NE.1.AND.LA1.NE.2)GOTO 1137
      IF(LA1.EQ.1)THEN
      WRITE(*,1139)
1139  FORMAT('0','Enter a valid areal source data base filename'/1X,
     *'with extension DAT such as B:FILE7.DAT in apostrophes  '\)
      READ(*,*)FNAME
      OPEN(6,FILE=FNAME,STATUS='NEW')
      DO 1142 J=1,NR
      DO 1141 I=1,NC
      WRITE(6,1140 )RH(I,J),R(I,J),CSCC(I,J)
1140  FORMAT(1X,F9.2,F12.2,F9.3)
1141  CONTINUE
1142  CONTINUE
      CLOSE (6)
      ENDIF
1143  ENDIF
1144  WRITE(*,1145)
```

```
 1145 FORMAT('0','Enter 1 to simulate local contaminant source'/1X,
     *'leakage through pit or basin bed with known contaminant'/1X,
     *'concentration into aquifer or aquifer layer. Enter 2 for'/1X,
     *'no such local source or to simulate a local source with'/1X,
     *'known contaminant load and unknown solute concentration.  '\)
      READ(*,*)NLC
      IF(NLC.NE.1.AND.NLC.NE.2)GOTO 1144
      IF(NLC.EQ.2)NPITS=0
      IF(NLC.EQ.1)THEN
      WRITE(*,1146)
 1146 FORMAT('0','Enter integer number of local'/1X,
     *'pits or basins (must be <51)  '\)
      READ(*,*)NPITS
      DO 1150 K=1,NPITS
      WRITE(*,1147)K
 1147 FORMAT(1X,'Enter real number vertical hydraulic'/1X,
     *'conductivity of pit or basin bed',I2,' usually'/1X,
     *'ranges from 1E-7 to 10 gpd/sq ft  '\)
      READ(*,*)PIT
      WRITE(*,1148)K
 1148 FORMAT(1X,'Enter real number thickness of'/1X,
     *'pit or basin bed',I2,' in ft  '\)
      READ(*,*)PTH
      PLEAK(K)=PIT/PTH
      WRITE(5,1149)K,PLEAK(K)
 1149 FORMAT(1X,'Leakance of pit or basin',I2,/1X,
     *' in gpd/cu ft=',F12.6)
 1150 CONTINUE
      ENDIF
 1152 WRITE(*,1153)
 1153 FORMAT('0','Enter 1 to simulate local contaminant source'/1X,
     *'rectangular in area with known contaminant load, unknown'/1X,
     *'solute concentration, and load injected directly into'/1X,
     *'aquifer or aquifer layer or 2 for no such source  '\)
      READ(*,*)NI1
      IF(NI1.NE.1.AND.NI1.NE.2)GOTO 1152
C
C     DEFINE STREAM SINK OR SOURCE NODE GEOMETRY AND
C     HYDRAULIC
C     CHARACTERISTICS
C
 1161 WRITE(*,1162)
 1162 FORMAT('0','Enter 1 to simulate surface water sinks at'/1X,
     *'nodes with known contaminant concentration, 2 to simulate'/1X,
     *'surface water sources at nodes, or 3 to continue'/1X,
     *'program  '\)
      READ(*,*)NN8
      IF(NN8.NE.1.AND.NN8.NE.2.AND.NN8.NE.3)GOTO 1161
      IF(NN8.EQ.3)NSSN=0
      IF(NN8.NE.3)THEN
      WRITE(*,1163)
 1163 FORMAT('0','Enter integer number of stream sink or'/1X,
```

```
     *'source nodes (must be < 101)  '\)
      READ(*,*)NSSN
      DO 1184 K=1,NSSN
      WRITE(*,1164)K
 1164 FORMAT(1X,'Stream sink or source node number=',I3)
      WRITE(5,1165)K
 1165 FORMAT(1X,'Stream sink or source number=',I3)
      WRITE(*,1166)
 1166 FORMAT(1X,'Enter integer I-coordinate of stream'/1X,
     *'sink or source node  '\)
      READ(*,*)ISS(K)
      WRITE(*,1167)
 1167 FORMAT(1X,'Enter integer J-coordinate of stream'/1X,
     *'sink or source node  '\)
      READ(*,*)JSS(K)
      I=ISS(K)
      J=JSS(K)
      WRITE(5,1168)I,J
 1168 FORMAT(1X,'I,J coordinates of stream node=',I3,',',I3)
      WRITE(*,1169)
 1169 FORMAT(1X,'Enter real number streambed width at'/1X,
     *'node in ft  '\)
      READ(*,*)STRW
      WRITE(5,1170)STRW
 1170 FORMAT(1X,'Streambed width at node in ft=',F9.2)
      WRITE(*,1171)
 1171 FORMAT(1X,'Enter real number streambed length at'/1X,
     *'node in ft  '\)
      READ(*,*)STRL
      WRITE(5,1172)STRL
 1172 FORMAT(1X,'Streambed length at node in ft=',F9.2)
      WRITE(*,1173)
 1173 FORMAT(1X,'Enter real number vertical hydraulic'/1X,
     *'conductivity of streambed at node in gpd/sq ft'/1X,
     *'usually ranges from .001 to 100 gpd/sq ft  '\)
      READ(*,*)PST
      WRITE(5,1174)PST
 1174 FORMAT(1X,'Vertical hydraulic conductivity of'/1X,
     *'streambed at node in gpd/sq ft=',F9.2)
      WRITE(*,1176)
 1176 FORMAT(1X,'Enter real number streambed thickness'/1X,
     *'at node, usually ranges from .1 to 10 ft  '\)
      READ(*,*)THS
      WRITE(5,1177)THS
 1177 FORMAT(1X,'Streambed thickness at node in ft=',F6.2)
      WRITE(*,1178)
 1178 FORMAT(1X,'Enter real number stream surface'/1X,
     *'elevation at node in ft  '\)
      READ(*,*)RH(I,J)
      WRITE(5,1179)RH(I,J)
 1179 FORMAT(1X,'Stream surface elevation at node in ft=',F9.2)
      RS(I,J)=(PST/THS)*STRW*STRL
```

```
      WRITE(*,1180)
 1180 FORMAT(1X,'Enter real number streambed base'/1X,
     *'elevation at node in ft  '\)
      READ(*,*)RD(I,J)
      WRITE(5,1181)RD(I,J)
 1181 FORMAT(1X,'Streambed base elevation at node'/1X,
     *'in ft=',F9.2)
      WRITE(*,1182)RS(I,J)
 1182 FORMAT(1X,'Simulation factor at node in gpd/ft=',F12.2)
      WRITE(5,1183)RS(I,J)
 1183 FORMAT(1X,'Simulation factor at node in gpd/ft=',F12.2)
      IF(NN8.EQ.1)MARK(I,J)=K
 1184 CONTINUE
      ENDIF
C
C     DEFINE PRODUCTION WELL SINK LOCATION AND STRENGTH
C
 1186 WRITE(*,1187)
 1187 FORMAT('0','Enter 1 to simulate production well sinks'/1X,
     *'or 2 for no production well sink. All production'/1X,
     *'well sinks must be located at nodes  '\)
      READ(*,*)NN7
      IF(NN7.NE.1.AND.NN7.NE.2)GOTO 1186
      IF(NN7.EQ.2)NPUMP=0
      IF(NN7.EQ.1)THEN
      WRITE(*,1200)
 1200 FORMAT('0','Enter integer number of production well sinks'/1X,
     *'(must be < 25)  '\)
      READ(*,*)NPUMP
      WRITE(5,1220)NPUMP
 1220 FORMAT(1X,'Number of production well sinks=',I3)
      DO 1280 K=1,NPUMP
      WRITE(*,1230)K
 1230 FORMAT(1X,'Production well sink number=',I3)
      WRITE(*,1240)
 1240 FORMAT(1X,'Enter integer I coordinate of production'/1X,
     *'well sink  '\)
      READ(*,*)IP(K)
      WRITE(*,1250)
 1250 FORMAT(1X,'Enter J coordinate of production'/1X,
     *'well sink  '\)
      READ(*,*)JP(K)
      WRITE(5,1260)K
 1260 FORMAT(1X,'Production well sink number=',I3)
      WRITE(5,1270)IP(K),JP(K)
 1270 FORMAT(1X,'I,J cordinates of production well'/1X,
     *'sink=',I3,',',I3)
      WRITE(*,1271)
 1271 FORMAT(1X,'Enter real number production well'/1X,
     *'discharge rate in gpm  '\)
      READ(*,*)Q1
      WRITE(5,1272)Q1
```

```
 1272 FORMAT(1X,'Production well discharge rate'/1X,
     *'in gpm=',F10.2)
      Q1=Q1*1440
      I=IP(K)
      J=JP(K)
      Q(I,J)=Q1
      MARK(I,J)=NSSN+K
      IF(NN8.EQ.2)MARK(I,J)=K
 1280 CONTINUE
      ENDIF
C
C     DEFINE AREAL LEAKAGE SINKS
C
 1316 WRITE(*,1318)
 1318 FORMAT('0','Enter 1 to simulate areal leakage'/1X,
     *'sinks or 2 for no areal leakage sinks. Areal leakage sinks'/1X,
     *'means areal contaminant migration from one aquifer'/1X,
     *'layer to an adjacent aquifer layer or areal contaminant'/1X,
     *'migration from one aquifer through an adjacent aquitard'/1X,
     *'and into another aquifer. There is contaminant mass'/1X,
     *'depletion in the aquifer or aquifer layer.  '\)
      READ(*,*)NLSK
      IF(NLSK.NE.1.AND.NLSK.NE.2)GOTO 1316
      IF(NLSK.EQ.2)NTT=0
      IF(NLSK.EQ.2)GOTO 1430
 1320 WRITE(*,1322)
 1322 FORMAT('0','Enter 1 to simulate two adjacent aquifer'/1X,
     *'layers or 2 to simulate aquifer-aquitard-sourcebed'/1X,
     *'system  '\)
      READ(*,*)NTT
      IF(NTT.NE.1.AND.NTT.NE.2)GOTO 1320
      IF(NTT.EQ.2)THEN
      WRITE(*,1324)
 1324 FORMAT('0','Enter real number default aquitard vertical'/1X,
     *'hydraulic conductivity in gpd/ sq ft, usually ranges'/1X,
     *'from 1E-5 to 10 gpd/sq ft  '\)
      READ(*,*)PSNK
      WRITE(*,1325)
 1325 FORMAT(1X,'Enter real number default aquitard thickness'/1X,
     *'in ft  '\)
      READ(*,*)TSNK
      DO 1327 J=1,NR
      DO 1326 I=1,NC
      P3(I,J)=PSNK
      THICK3(I,J)=TSNK
 1326 CONTINUE
 1327 CONTINUE
 1328 WRITE(*,1329)
 1329 FORMAT('0','Enter 1 for keyboard sourcebed head data'/1X,
     *'base input or 2 to import ASCII sourcebed head data'/1X,
     *'base file previously created with this program  '\)
      READ(*,*)KSD
```

```
      IF(KSD.NE.1.AND.KSD.NE.2)GOTO 1328
      IF(KSD.EQ.1)GOTO 1340
      WRITE(*,1330)
 1330 FORMAT('0','Enter a valid sourcebed head filename'/1X,
     *'with extension DAT such as B:FILE3.DAT in'/1X,
     *'apostrophes  '\)
      READ(*,*)FNAME
      OPEN(6,FILE=FNAME)
      DO 1333 J=1,NR
      DO 1332 I=1,NC
      READ(6,1331)H1(I,J)
 1331 FORMAT(1X,F9.2)
 1332 CONTINUE
 1333 CONTINUE
      CLOSE (6)
      GOTO 1351
 1340 DO 1344 J=1,NR
      DO 1343 I=1,NC
      WRITE(*,1341)I,J
 1341 FORMAT('0','I,J coordinates of node=',I3,',',I3)
      WRITE(*,1342)
 1342 FORMAT(1X,'Enter real number sourcebed head elevation'/1X,
     *'in ft  '\)
      READ(*,*)H1(I,J)
 1343 CONTINUE
 1344 CONTINUE
 1345 WRITE(*,1346)
 1346 FORMAT('0','Enter 1 to export sourcebed head'/1X,
     *'data base to a file or 2 for no file  '\)
      READ(*,*)NEX
      IF(NEX.NE.1.AND.NEX.NE.2)GOTO 1345
      IF(NEX.EQ.1)THEN
      WRITE(*,1347)
 1347 FORMAT('0','Enter a valid sourcebed head'/1X,
     *'filename with extension DAT such as B:FILE4.DAT'/1X,
     *'in apostrophes  '\)
      READ(*,*)FNAME
      OPEN(6,FILE=FNAME,STATUS='NEW')
      DO 1350 J=1,NR
      DO 1349 I=1,NC
      WRITE(6,1348)H1(I,J)
 1348 FORMAT(1X,F9.2)
 1349 CONTINUE
 1350 CONTINUE
      CLOSE (6)
      ENDIF
 1351 GOTO 1382
      ENDIF
 1352 WRITE(*,1353)
 1353 FORMAT(1X,'Enter real number default aquifer layer 1'/1X,
     *'vertical hydraulic conductivity in gpd/sq ft. Aquifer layer'/1X,
     *'1 is the aquifer layer under investigation and aquifer'/1X,
```

```
     *'layer 2 is the adjoining aquifer layer. Usually ranges from'/1X,
     *'0.1 to 1000 gpd/sq ft  '\)
      READ(*,*)PV1
      WRITE(*,1354)
1354 FORMAT(1X,'Enter real number default aquifer layer 2'/1X,
     *'vertical hydraulic conductivity in gpd/sq ft, usually'/1X,
     *'ranges from 0.1 to 1000 gpd/sq ft  '\)
      READ(*,*)PV2
      WRITE(*,1355)
1355 FORMAT(1X,'Enter real number default aquifer l'/1X,
     *'thickness in ft  '\)
      READ(*,*)MA1
      WRITE(*,1356)
1356 FORMAT(1X,'Enter real number default aquifer layer 2'/1X,
     *'thickness in ft  '\)
      READ(*,*)MA2
      DO 1358 J=1,NR
      DO 1357 I=1,NC
      P4(I,J)=PV1
      P5(I,J)=PV2
      THICK4(I,J)=MA1
      THICK5(I,J)=MA2
1357 CONTINUE
1358 CONTINUE
1360 WRITE(*,1361)
1361 FORMAT('0','Enter 1 for keyboard aquifer layer 2 head'/1X,
     *'data base input or 2 to import ASCII aquifer layer 2'/1X,
     *'data base file previously created with this program  '\)
      READ(*,*)KSD
      IF(KSD.NE.1.AND.KSD.NE.2)GOTO 1360
      IF(KSD.EQ.1)GOTO 1366
      WRITE(*,1362)
1362 FORMAT('0','Enter a valid aquifer layer 2 head filename'/1X,
     *'with extension DAT such as B:FILE3.DAT in'/1X,
     *'apostrophes  '\)
      READ(*,*)FNAME
      OPEN(6,FILE=FNAME)
      DO 1365 J=1,NR
      DO 1364 I=1,NC
      READ(6,1363)H2(I,J)
1363 FORMAT(1X,F9.2)
1364 CONTINUE
1365 CONTINUE
      CLOSE (6)
      GOTO 1382
1366 DO 1373 J=1,NR
      DO 1372 I=1,NC
      WRITE(*,1370)I,J
1370 FORMAT('0','I,J coordinates of node=',I3,',',I3)
      WRITE(*,1371)
1371 FORMAT(1X,'Enter real number aquifer layer 2 head'/1X,
     *'elevation in ft  '\)
```

```
      READ(*,*)H2(I,J)
 1372 CONTINUE
 1373 CONTINUE
 1375 WRITE(*,1376)
 1376 FORMAT('0','Enter 1 to export aquifer layer 2 head'/1X,
     *'data base to a file or 2 for no file  '\)
      READ(*,*)NEX
      IF(NEX.NE.1.AND.NEX.NE.2)GOTO 1375
      IF(NEX.EQ.1)THEN
      WRITE(*,1377)
 1377 FORMAT('0','Enter a valid aquifer layer 2 head'/1X,
     *'filename with extension DAT such as B:FILE4.DAT'/1X,
     *'in apostrophes  '\)
      READ(*,*)FNAME
      OPEN(6,FILE=FNAME,STATUS='NEW')
      DO 1380 J=1,NR
      DO 1379 I=1,NC
      WRITE(6,1378)H2(I,J)
 1378 FORMAT(1X,F9.2)
 1379 CONTINUE
 1380 CONTINUE
      CLOSE (6)
      ENDIF
 1382 IF(NTT.EQ.0.OR.NTT.EQ.1)NCS=0
      IF(NTT.EQ.2)THEN
      RD1V=1.0
 1383 WRITE(*,1384)
 1384 FORMAT('0','Enter 1 to simulate contaminant sorption'/1X,
     *'in aquitard or 2 for no sorption  '\)
      READ(*,*)NCS
      IF(NCS.NE.1.AND.NCS.NE.2)GOTO 1383
      IF(NCS.EQ.2)GOTO 1390
      WRITE(*,1385)
 1385 FORMAT('0','It is assumed that sorption reactions'/1X,
     *'are fast and reversible and the isotherm is linear.'/1X,
     *'It is further assumed that there is only one'/1X,
     *'contaminant in solution and no mixture of contaminants.')
      WRITE(*,1386)
 1386 FORMAT('0','Enter real number distribution coefficient'/1X,
     *'for aquitard in ml/g, usually ranges from 1E-5 to'/1X,
     *'1E6 ml/g  '\)
      READ(*,*)DKL
      WRITE(*,1387)
 1387 FORMAT(1X,'Enter real number bulk mass density of dry'/1X,
     *'aquitard skeleton in g/cu cm, usually ranges from 0.7'/1X,
     *'to 3.2 g/cu cm  '\)
      READ(*,*)RHOL
      RD1V=1+(RHOL/EPOR)*DKL
      WRITE(5,1388)RD1V
 1388 FORMAT('0','Aquitard retardation factor=',F9.2)
      WRITE(*,1389)RD1V
 1389 FORMAT('0','Aquitard retardation factor=',F9.2)
```

```
      CLOSE (5)
1390 ENDIF
1391 WRITE(*,1392)
1392 FORMAT('0','Enter 1 to define aquitard or aquifer'/1X,
    *'layer hydraulic characteristic nodal values differing'/1X,
    *'from default values or 2 to continue program  '\)
      READ(*,*)KDF
      IF(KDF.NE.1.AND.KDF.NE.2)GOTO 1391
      IF(KDF.EQ.1)THEN
1393 WRITE(*,1394)
1394 FORMAT('0','Enter integer number of grid blocks'/1X,
    *'containing nodes whose default values are to be'/1X,
    *'changed  '\)
      READ(*,*)NGB
      DO 1426 L=1,NGB
      WRITE(*,1395)L
1395 FORMAT('0','Grid block number=',I3)
      WRITE(*,1396)
1396 FORMAT(1X,'Enter integer grid block left- and right-'/1X,
    *'most column I coordinates separated by a comma  '\)
      READ(*,*)IVG1,IVG2
      WRITE(*,1397)
1397 FORMAT(1X,'Enter integer grid block upper- and lower-'/1X,
    *'most row J coordinates separated by a comma  '\)
      READ(*,*)JVG1,JVG2
      DO 1424 J=JVG1,JVG2
      DO 1422 I=IVG1,IVG2
      WRITE(*,1398)I
1398 FORMAT(1X,'I-coordinate of node=',I3)
      WRITE(*,1400)J
1400 FORMAT(1X,'J-coordinate of node=',I3)
      IF(NTT.EQ.2)THEN
      WRITE(*,1402)
1402 FORMAT(1X,'Enter real number aquitard hydraulic'/1X,
    *'conductivity at node in gpd/sq ft  '\)
      READ(*,*)P3(I,J)
      WRITE(*,1404)
1404 FORMAT(1X,'Enter real number aquitard thickness'/1X,
    *'at node in ft  '\)
      READ(*,*)THICK3(I,J)
      GOTO 1422
      ENDIF
      WRITE(*,1406)
1406 FORMAT(1X,'Enter real number aquifer layer 1'/1X,
    *'hydraulic conductivity in gpd/sq ft  '\)
      READ(*,*)P4(I,J)
      WRITE(*,1408)
1408 FORMAT(1X,'Enter real number aquifer layer 2'/1X,
    *'hydraulic conductivity in gpd/sq ft  '\)
      READ(*,*)P5(I,J)
      WRITE(*,1410)
1410 FORMAT(1X,'Enter real number aquifer layer 1'/1X,
```

```
     *'thickness in ft  '\)
      READ(*,*)THICK4(I,J)
      WRITE(*,1412)
1412 FORMAT(1X,'Enter real number aquifer layer 2'/1X,
     *'thickness in ft  '\)
      READ(*,*)THICK4(I,J)
1422 CONTINUE
1424 CONTINUE
1426 CONTINUE
      ENDIF
1430 CLOSE (5)
      END
```

```
C
      SUBROUTINE CONTIN(MFL)
C
C     THIS SUBROUTINE GENERATES PARTICLES TO
C     SIMULATE RECTANGULAR CONTAMINANT SOURCE.
C     THE SOURCE AREA GEOMETRY IS DEFINED AND
C     PARTICLES ARE DISTRIBUTED RANDOMLY WITHIN
C     THE AREA AND IN TIME
C
C     METACOMMANDS
C
$LARGE
C
C     DECLARATIONS
C
      COMMON/NSUB7/NP,NPART(50,50)
      COMMON/NSUB6/MAXP,NSORS,NIJ,NSSN,MARK(50,50),NRAD
      COMMON/AREA7/PM,DISPL,DISPT,EPOR,APOR
      COMMON/AREA6/DELTA,THICKC(50,50),PS(50,50)
      COMMON/AREA5/X(5001),Y(5001),V(50,50,2)
      COMMON/AREA8/TMAP,ANC,ANR,TABLE(124)
      COMMON/NSUB1/NC,NR
      COMMON/AREA3/RD1,HL,PLEAK(50),Q(50,50),RD1V
      COMMON/AREA2/DELX(51),DELY(51),P1(50,50),P2(50,50)
C
      OPEN(5,FILE='LPT1:')
      WRITE(*,42)
   42 FORMAT('0','Continuous leakage of contaminants'/1X,
     *'with known contaminant loads from landfills, pits,'/1X,
     *'basins, ditches, streams, lakes, etc. may be'/1X,
     *'simulated')
      WRITE(*,44)
   44 FORMAT('0','Enter integer number of rectangular'/1X,
     *'source areas (must be < 51)  '\)
      READ(*,*)NSA
      WRITE(5,46)NSA
   46 FORMAT('0','Number of rectangular areas=',I3)
      DO 115 K=1,NSA
      WRITE(*,47)K
   47 FORMAT('0','Rectangular source area number=',I3)
      WRITE(5,48)K
   48 FORMAT('0','Rectangular source area number=',I3)
      WRITE(*,50)
   50 FORMAT('0','Enter total contaminant load during'/1X,
     *'this time increment in pounds  '\)
      READ(*,*)PL
```

```
      WRITE(5,60)PL
   60 FORMAT('0','Total contaminant load during this'/1X,
     *'time increment in pounds=',F10.3)
      PT1=PL/PM
      NPT1=ANINT(PT1)
      PL=NPT1*PM
      WRITE(*,70)
   70 FORMAT(1X,'Enter real number I and J coordinates'/1X,
     *'of upper left corner of rectangular source area which'/1X,
     *'may be a point, line or square by proper choice of'/1X,
     *'area corner coordinates and dimensions of area.'/1X,
     *'Separate coordinates by comma.  '\)
      READ(*,*)X1,Y1
      WRITE(5,80)X1,Y1
   80 FORMAT(1X,'I,J coordinates of upper left corner'/1X,
     *'of rectangular source area=',F5.2,',',F5.2)
      WRITE(*,90)
   90 FORMAT(1X,'Enter real number I and J coordinates of'/1X,
     *'lower right corner of rectangular source area'/1X,
     *'separated by a comma.  '\)
      READ(*,*)X2,Y2
      WRITE(5,100)X2,Y2
  100 FORMAT(1X,'I,J coordinates of lower right corner'/1X,
     *'of rectangular source area=',F5.2,',',F5.2)
      IF(MFL.EQ.1)WRITE(*,102)
  102 FORMAT('0','COMPUTATIONS ARE IN PROGRESS')
      MFL=2
      DX=ABS(X1-X2)
      DY=ABS(Y1-Y2)
      P=PL
  110 IF(RAN(0).GT.P/PM)GOTO 115
      P=P-PM
      XX=X1+DX*RAN(0)
      YY=Y1+DY*RAN(0)
      DE=DELTA*RAN(0)
      IF(MOVE(XX,YY,DE).EQ.0)CALL ADD(XX,YY)
      GO TO 110
  115 CONTINUE
  130 CLOSE (5)
      END
```

```
C
      SUBROUTINE DATPRN(HHE,MCX)
C
C      THIS SUBROUTINE PRINTS A HARD COPY OF
C      VARIABLE VALUES AS REQUESTED
C
C      METACOMMANDS
C
$LARGE
C
C      DECLARATIONS
C
      DIMENSION HHE(50,50)
      COMMON/NSUB1/NC,NR
C
      OPEN(5,FILE='LPT1:')
      N1=1
      N2=8
      IF(NC.LT.N2)N2=NC
  110 WRITE(5,120)(I,I=N1,N2)
  120 FORMAT(3X,8I9)
      DO 140 J=1,NR
      IF(MCX.EQ.2)THEN
      WRITE(5,122)J,(HHE(I,J),I=N1,N2)
  122 FORMAT(I3,8F9.4)
      GOTO 140
      ENDIF
      IF(MCX.EQ.3)THEN
      WRITE(5,124)J,(HHE(I,J),I=N1,N2)
  124 FORMAT(I3,8F9.6)
      GOTO 140
      ENDIF
      IF(MCX.EQ.4)THEN
      WRITE(5,125)J,(HHE(I,J),I=N1,N2)
  125 FORMAT(I3,8F9.3)
      GOTO 140
      ENDIF
      WRITE(5,126)J,(HHE(I,J),I=N1,N2)
  126 FORMAT(I3,8F9.2)
  140 CONTINUE
      IF(N2.EQ.NC)GOTO 150
      N1=N1+8
      N2=N2+8
      IF(NC.LT.N2)N2=NC
      GOTO 110
  150 CLOSE (5)
      END
```

```
C
      SUBROUTINE DISPER
C
C     THIS SUBROUTINE DEFINES DISPERSIVITY
C
C     METACOMMANDS
C
$LARGE
C
C     DECLARATIONS
C
      COMMON/AREA5/X(5001),Y(5001),V(50,50,2)
      COMMON/AREA7/PM,DISPL,DISPT,EPOR,APOR
C
      OPEN(5,FILE='LPT1:')
   10 WRITE(*,20)
   20 FORMAT('0','Dispersivity tends to increase and to'/1X,
     *'approach some asymptotic value (usually assumed'/1X,
     *'to be 300 ft) with expansions of contaminant'/1X,
     *'plumes. Dispersivity may be estimated as a function'/1X,
     *'of the mean contaminant particle travel distance'/1X,
     *'which is related to groundwater flow velocities.'/1X,
     *'Longitudinal dispersivity is usually assumed to'/1X,
     *'be 1/10 of the length of the contaminant plume up'/1X,
     *'to 300 ft. Dispersivity changes with time. Enter 1'/1X,
     *'for display of grid groundwater velocities to assist'/1X,
     *'estimation of dispersivity or 2 for no display  '\)
      READ(*,*)NDIS
      IF(NDIS.NE.1.AND.NDIS.NE.2)GOTO 10
      IF(NDIS.EQ.1)THEN
   25 WRITE(*,30)
   30 FORMAT('0','Enter I and J coordinates of node where'/1X,
     *'groundwater velocities are desired separated by'/1X,
     *'a comma  '\)
      READ(*,*)I,J
      WRITE(*,40)V(I,J,2)
   40 FORMAT(1X,'Groundwater velocity in x-direction in'/1X,
     *'ft/day=',F9.3)
      WRITE(*,50)V(I,J,1)
   50 FORMAT(1X,'Groundwater velocity in y-direction'/1X,
     *'in ft/day=',F9.3)
   52 WRITE(*,54)
   54 FORMAT('0','Enter 1 for velocity values at another'/1X,
     *'node or 2 to continue program  '\)
      READ(*,*)NVN
      IF(NVN.NE.1.AND.NVN.NE.2)GOTO 52
```

```
      IF(NVN.EQ.1)GOTO 25
      ENDIF
      WRITE(*,60)
   60 FORMAT('0','Enter real number longitudinal dispersivity'/1X,
     *'in ft without sorption, usually ranges from .1 to'/1X,
     *'300 ft  '\)
      READ(*,*)DISPL
      WRITE(5,70)DISPL
   70 FORMAT(1X,'Longitudinal dispersivity in ft=',F8.3)
      WRITE(*,80)
   80 FORMAT(1X,'Enter real number transverse dispersity in ft'/1X,
     *'without sorption, usually ranges from 1/5 to 1/10'/1X,
     *'of longitudinal dispersity  '\)
      READ(*,*)DISPT
      WRITE(5,90)DISPT
   90 FORMAT(1X,'Transverse dispersivity in ft=',F8.3)
      CLOSE (5)
      END
```

```
C
      SUBROUTINE FILEI1(NN8)
C
C     THIS SUBROUTINE IMPORTS AQUIFER GEOMETRY
C     AND CHARACTERISTIC DATA BASE
C
C     METACOMMANDS
C
$LARGE
C
C     DECLARATIONS
C
      CHARACTER*14FNAME
      COMMON/NSUB1/NC,NR
      COMMON/AREA2/DELX(51),DELY(51),P1(50,50),P2(50,50)
      COMMON/AREA6/DELTA,THICKC(50,50),PS(50,50)
      COMMON/ASUB2/CH(50,50),BOT(50,50),H(50,50)
      COMMON/AREA1/RH(50,50),R(50,50),RD(50,50),RS(50,50)
      COMMON/AREA7/PM,DISPL,DISPT,EPOR,APOR
      COMMON/NSUB2/NPUMP,NPITS
      COMMON/NSUB4/IP(24),JP(24),ISS(100),JSS(100)
      COMMON/NSUB6/MAXP,NSORS,NIJ,NSSN,MARK(50,50),NRAD
      COMMON/AREA3/RD1,HL,PLEAK(50),Q(50,50),RD1V
      COMMON/AREA5/X(5001),Y(5001),V(50,50,2)
      COMMON/NSUB3/IJJ(24),JIJ(24)
      COMMON/ASUB1/CONSNK(50,50),CONC(50,50),
     *CSCC(50,50),CONSOR(50,50)
      COMMON/ASUB3/P3(50,50),P4(50,50),P5(50,50),THICK3(50,50)
      COMMON/ASUB4/THICK4(50,50),THICK5(50,50),H1(50,50),H2(50,50)
      COMMON/NSUB8/NAL,NLSK,NTT,NPE,NN1,IWS,NLC,NI1
      COMMON/NSUB9/LEK,NRE,NCS
      COMMON/NSUB7/NP,NPT(50,50)
      COMMON/AREA8/TMAP,ANC,ANR,TABLE(124)
   10 WRITE(*,20)
   20 FORMAT('0','Enter a valid data base specification'/1X,
     *'filename with extension DAT of ASCII file to be'/1X,
     *'imported such as B:FILE4.DAT in apostrophes  '\)
      READ(*,*)FNAME
      OPEN(6,FILE=FNAME)
      READ(6,30)NC,NR,NSORS,NPUMP,NSSN,NPITS,NIJ,LEK
   30 FORMAT(1X,8I3)
      READ(6,32)NAL,NLSK,NTT,NPE,NN1,IWS,NLC,NI1,NN8
   32 FORMAT(1X,9I3)
      READ(6,40)EPOR,APOR,RD1,HL,PM,RD1V
   40 FORMAT(1X,2F7.4,F9.2,F15.2,F10.3,F9.2)
      READ(6,41)ANC,ANR,TMAP,NP
```

```
 41 FORMAT(1X,2F6.2,F9.3,I3)
    READ(6,42)MAXP
 42 FORMAT(1X,I5)
    DO 80 I=1,NPUMP
    READ(6,70)IP(I),JP(I)
 70 FORMAT(1X,2I3)
 80 CONTINUE
    DO 104 I=1,NSSN
    READ(6,102)ISS(I),JSS(I)
102 FORMAT(1X,2I3)
104 CONTINUE
    DO 108 I=1,NIJ
    READ(6,106)IJJ(I),JIJ(I)
106 FORMAT(1X,2I3)
108 CONTINUE
    DO 140 J=1,NR
    DO 130 I=1,NC
    READ(6,110)P1(I,J),P2(I,J),CH(I,J),BOT(I,J)
110 FORMAT(1X,4F9.2)
    READ(6,112)P3(I,J),P4(I,J),P5(I,J),H1(I,J),H2(I,J),Q(I,J)
112 FORMAT(1X,F12.6,4F9.2,F12.2)
    READ(6,114)THICK3(I,J),THICK4(I,J),THICK5(I,J),PS(I,J)
114 FORMAT(1X,3F9.2,F10.3)
    READ(6,120)V(I,J,1),V(I,J,2),MARK(I,J)
120 FORMAT(1X,2F9.3,I4)
    READ(6,122)R(I,J),RH(I,J),THICKC(I,J)
122 FORMAT(1X,F12.2,2F9.2)
    READ(6,124)CONSOR(I,J),CSCC(I,J)
124 FORMAT(1X,2F10.3)
130 CONTINUE
140 CONTINUE
    DO 144 I=1,NPITS
    READ(6,142)PLEAK(I)
142 FORMAT(1X,F12.6)
144 CONTINUE
    DO 170 I=1,NC
    READ(6,160)DELX(I)
160 FORMAT(1X,F9.2)
170 CONTINUE
    DO 190 J=1,NR
    READ(6,180)DELY(J)
180 FORMAT(1X,F9.2)
190 CONTINUE
    DO 210 K=1,NSSN
    I=ISS(K)
    J=JSS(K)
    READ(6,200)RD(I,J),RS(I,J)
200 FORMAT(1X,F9.2,F10.3)
210 CONTINUE
    READ(6,212)DELTA
212 FORMAT(1X,F9.3)
    CLOSE (6)
    END
```

```
C
      SUBROUTINE FILEI2
C
C
C     THIS SUBROUTINE IMPORTS AQUIFER HEAD DATA BASE
C
C     METACOMMANDS
C
$LARGE
C
C     DECLARATIONS
C
      CHARACTER*14FNAME
      COMMON/NSUB1/NC,NR
      COMMON/ASUB2/CH(50,50),BOT(50,50),H(50,50)
   10 WRITE(*,20)
   20 FORMAT('0','Enter a valid head data base specification'/1X,
     *'filename with extension DAT of ASCII file to be'/1X,
     *'imported such as B:FILE4.DAT in apostrophes   '\)
      READ(*,*)FNAME
      OPEN(6,FILE=FNAME)
      DO 140 J=1,NR
      DO 130 I=1,NC
      READ(6,110)H(I,J)
  110 FORMAT(1X,F9.2)
  130 CONTINUE
  140 CONTINUE
      CLOSE (6)
      END
```

```
C
      SUBROUTINE FILEI3
C
C     THIS SUBROUTINE IMPORTS PRE-EXISTING CONTAMINANT
C     CONCENTRATION DATA BASE
C
C     METACOMMANDS
C
$LARGE
C
C     DECLARATIONS
C
      CHARACTER*14FNAME
      COMMON/NSUB1/NC,NR
      COMMON/ASUB1/CONSNK(50,50),CONC(50,50),
     *CSCC(50,50),CONSOR(50,50)
   10 WRITE(*,20)
   20 FORMAT('0','Enter a pre-existing contaminant condition'/1X,
     *'valid data base filename with extension DAT of ASCII'/1X,
     *'file to be imported such as B:FILE4.DAT in apostrophes  '\)
      READ(*,*)FNAME
      OPEN(6,FILE=FNAME)
      DO 140 J=1,NR
      DO 130 I=1,NC
      READ(6,110)CONSOR(I,J)
  110 FORMAT(1X,F10.3)
  130 CONTINUE
  140 CONTINUE
      CLOSE (6)
      END
```

```
C
      SUBROUTINE FILEO1(NN8)
C
C     THIS SUBROUTINE EXPORTS AQUIFER GEOMETRY AND
C     CHARACTERISTICS DATA BASE
C
C     METACOMMANDS
C
$LARGE
C
C     DECLARATIONS
C
      CHARACTER*14FNAME
      COMMON/NSUB1/NC,NR
      COMMON/AREA2/DELX(51),DELY(51),P1(50,50),P2(50,50)
      COMMON/AREA6/DELTA,THICKC(50,50),PS(50,50)
      COMMON/ASUB2/CH(50,50),BOT(50,50),H(50,50)
      COMMON/AREA1/RH(50,50),R(50,50),RD(50,50),RS(50,50)
      COMMON/AREA7/PM,DISPL,DISPT,EPOR,APOR
      COMMON/NSUB2/NPUMP,NPITS
      COMMON/NSUB4/IP(24),JP(24),ISS(100),JSS(100)
      COMMON/NSUB6/MAXP,NSORS,NIJ,NSSN,MARK(50,50),NRAD
      COMMON/AREA3/RD1,HL,PLEAK(50),Q(50,50),RD1V
      COMMON/AREA5/X(5001),Y(5001),V(50,50,2)
      COMMON/NSUB3/IJJ(24),JIJ(24)
      COMMON/ASUB1/CONSNK(50,50),CONC(50,50),CSCC(50,50),
     *CONSOR(50,50)
      COMMON/ASUB3/P3(50,50),P4(50,50),P5(50,50),THICK3(50,50)
      COMMON/ASUB4/THICK4(50,50),THICK5(50,50),H1(50,50),H2(50,50)
      COMMON/NSUB8/NAL,NLSK,NTT,NPE,NN1,IWS,NLC,NI1
      COMMON/NSUB9/LEK,NRE,NCS
      COMMON/NSUB7/NP,NPT(50,50)
      COMMON/AREA8/TMAP,ANC,ANR,TABLE(124)
   10 WRITE(*,20)
   20 FORMAT('0','Enter a monitor well, sink, grid, hydraulic'/1X,
     *'characteristic valid data base filename with extension'/1X,
     *'DAT of ASCII file to be exported such as B:FILE1.DAT'/1X,
     *'in apostrophes  '\)
      READ(*,*)FNAME
      OPEN(6,FILE=FNAME,STATUS='NEW')
      WRITE(6,30)NC,NR,NSORS,NPUMP,NSSN,NPITS,NIJ,LEK
   30 FORMAT(1X,8I3)
      WRITE(6,32)NAL,NLSK,NTT,NPE,NN1,IWS,NLC,NI1,NN8
   32 FORMAT(1X,9I3)
      WRITE(6,40)EPOR,APOR,RD1,HL,PM,RD1V
   40 FORMAT(1X,2F7.4,F9.2,F15.2,F10.3,F9.2)
      WRITE(6,41)ANC,ANR,TMAP,NP
```

```
 41 FORMAT(1X,2F6.2,F9.3,I3)
    WRITE(6,42)MAXP
 42 FORMAT(1X,I5)
    DO 80 I=1,NPUMP
    WRITE(6,70)IP(I),JP(I)
 70 FORMAT(1X,2I3)
 80 CONTINUE
    DO 104 I=1,NSSN
    WRITE(6,102)ISS(I),JSS(I)
102 FORMAT(1X,2I3)
104 CONTINUE
    DO 108 I=1,NIJ
    WRITE(6,106)IJJ(I),JIJ(I)
106 FORMAT(1X,2I3)
108 CONTINUE
    DO 140 J=1,NR
    DO 130 I=1,NC
    WRITE(6,110)P1(I,J),P2(I,J),CH(I,J),BOT(I,J)
110 FORMAT(1X,4F9.2)
    WRITE(6,112)P3(I,J),P4(I,J),P5(I,J),H1(I,J),H2(I,J),Q(I,J)
112 FORMAT(1X,F12.6,4F9.2,F12.2)
    WRITE(6,114)THICK3(I,J),THICK4(I,J),THICK5(I,J),PS(I,J)
114 FORMAT(1X,3F9.2,F10.3)
    WRITE(6,120)V(I,J,1),V(I,J,2),MARK(I,J)
120 FORMAT(1X,2F9.3,I4)
    WRITE(6,122)R(I,J),RH(I,J),THICKC(I,J)
122 FORMAT(1X,F12.2,2F9.2)
    WRITE(6,124)CONSOR(I,J),CSCC(I,J)
124 FORMAT(1X,2F10.3)
130 CONTINUE
140 CONTINUE
    DO 144 I=1,NPITS
    WRITE(6,142)PLEAK(I)
142 FORMAT(1X,F12.6)
144 CONTINUE
    DO 170 I=1,NC
    WRITE(6,160)DELX(I)
160 FORMAT(1X,F9.2)
170 CONTINUE
    DO 190 J=1,NR
    WRITE(6,180)DELY(J)
180 FORMAT(1X,F9.2)
190 CONTINUE
    DO 210 K=1,NSSN
    I=ISS(K)
    J=JSS(K)
    WRITE(6,200)RD(I,J),RS(I,J)
200 FORMAT(1X,F9.2,F10.3)
210 CONTINUE
    WRITE(6,212)DELTA
212 FORMAT(1X,F9.3)
    CLOSE (6)
    END
```

```
C
      SUBROUTINE FILEO2
C
C
C     THIS SUBROUTINE EXPORTS AQUIFER HEAD DATA BASE
C
C     METACOMMANDS
C
$LARGE
C
C     DECLARATIONS
C
      CHARACTER*14FNAME
      COMMON/NSUB1/NC,NR
      COMMON/ASUB2/CH(50,50),BOT(50,50),H(50,50)
   10 WRITE(*,20)
   20 FORMAT('0','Enter a valid head data base specification'/1X,
     *'filename with extension DAT of ASCII file to be'/1X,
     *'exported such as B:FILE4.DAT in apostrophes  '\)
      READ(*,*)FNAME
      OPEN(6,FILE=FNAME,STATUS='NEW')
      DO 140 J=1,NR
      DO 130 I=1,NC
      WRITE(6,110)H(I,J)
  110 FORMAT(1X,F9.2)
  130 CONTINUE
  140 CONTINUE
      CLOSE (6)
      END
```

```
C
      SUBROUTINE FILEO3
C
C     THIS SUBROUTINE EXPORTS PRE-EXISTING CONTAMINANT
C     CONCENTRATION DATA BASE
C
C     METACOMMANDS
C
$LARGE
C
C     DECLARATIONS
C
      CHARACTER*14FNAME
      COMMON/NSUB1/NC,NR
      COMMON/ASUB1/CONSNK(50,50),CONC(50,50),
     *CSCC(50,50),CONSOR(50,50)
   10 WRITE(*,20)
   20 FORMAT('0','Enter a pre-existing contaminant condition'/1X,
     *'valid data base filename with extension DAT of ASCII'/1X,
     *'file to be exported such as B:FILE4.DAT in apostrophes  '\)
      READ(*,*)FNAME
      OPEN(6,FILE=FNAME,STATUS='NEW')
      DO 140 J=1,NR
      DO 130 I=1,NC
      WRITE(6,110)CONSOR(I,J)
  110 FORMAT(1X,F10.3)
  130 CONTINUE
  140 CONTINUE
      CLOSE (6)
      END
```

```
C
      SUBROUTINE HEAD(H,GS,PP)
C
C     THIS PROGRAM CREATES A UNIFORM GRADIENT
C     IN THE X-DIRECTION HEAD DATA BASE
C
C
C     DECLARATIONS
C
      CHARACTER*14FNAME
      DIMENSION H(50,50)
      COMMON/NSUB1/NC,NR
      COMMON/AREA7/PM,DISPL,DISPT,EPOR,APOR
C
      WRITE(*,10)
   10 FORMAT('0','Uniform hydraulic conductivity, effective'/1X,
     *'porosity, and grid spacing is assumed. Number of columns'/1X,
     *'and rows, effective porosity, hydraulic conductivity'/1X,
     *'and grid spacing are those that were specified previously'/1X,
     *'in this program.')
      WRITE(*,25)
   25 FORMAT(1X,'Enter real number elevation of water'/1X,
     *'level along column 1 in ft  '\)
      READ(*,*)EL
      WRITE(*,60)
   60 FORMAT(1X,'Enter real number groundwater seepage velocity'/1X,
     *'in ft/day  '\)
      READ(*,*)VEL
      DELTAH=VEL*7.48*GS*EPOR/PP
      DO 80 J=1,NR
      DO 70 I=1,NC
      IF(I.EQ.1)THEN
      H(I,J)=EL
      GOTO 70
      ENDIF
      H(I,J)=H(I-1,J)-DELTAH
   70 CONTINUE
   80 CONTINUE
   82 WRITE(*,84)
   84 FORMAT('0','Enter 1 to create a head data base file'/1X,
     *'or 2 for no file  '\)
      READ(*,*)K11
      IF(K11.NE.1.AND.K11.NE.2)GOTO 82
      IF(K11.EQ.1)THEN
      WRITE(*,90)
   90 FORMAT(1X,'Enter a valid head data base filename with'/1X,
```

```
   *'extension DAT such as B:FILE1.DAT in apostrophes   '\)
    READ(*,*)FNAME
    OPEN(6,FILE=FNAME,STATUS='NEW')
    DO 110 J=1,NR
    DO 100 I=1,NC
    WRITE(6,95)H(I,J)
 95 FORMAT(1X,F9.2)
100 CONTINUE
110 CONTINUE
    CLOSE (6)
    ENDIF
    END
```

```
C
      SUBROUTINE INFOM
C
C     THIS SUBROUTINE PROVIDES PROGRAM INFORMATION
C
      WRITE(*,10)
   10 FORMAT(1X,
     *'Program:       GWTR3D'/1X,
     *'Author:        W.C.Walton modified program presented'/1X,
     *'               by Prickett, et al. 1981.'/1X,
     *'Date:          August 1988'/1X,
     *'Version:       IBM/PC Microsoft FORTRAN77 v3.20 or 4.00'/1X,
     *'               Copyright 1988 Lewis Publishers, Inc.'/1X
     *'               IBM/PC is a trademark of International Business'/1X,
     *'               Machines Corporation'/1X,
     *'               Microsoft is a trademark of Microsoft Corp.'/1X,
     *'Description:   Modular program for a one, two, or three'/1X,
     *'               dimensional discrete parcel random walk'/1X,
     *'               model simulating single contaminant'/1X,
     *'               transport with constant density, dispersion,'/1X,
     *'               sorption, and radioactive decay in saturated'/1X,
     *'               aquifer systems. Hydraulic characteristics'/1X,
     *'               may vary horizontally and vertically.')
      PAUSE
      WRITE(*,20)
   20 FORMAT('0',
     *'References:    Ahlstrom,S.W.,H.P.Foote,R.C.Arnett,C.R.Cole,'/1X,
     *'               R.J.Serne. 1977. Multicomponent mass transport'/1X,
     *'               model. Theory and numerical implementation'/1X,
     *'               (discrete parcel random walk version). Battelle'/1X,
     *'               Rept. BNWL 2127.'/1X,
     *'               Prickett,T.A.,T.G.Naymik, and C.G.Lonnquist.'/1X,
     *'               1981. A Random-Walk solute transport model for'/1X,
     *'               selected groundwater quality evaluations.'/1X,
     *'               Illinois State Water Survey. Bull. 65. 103 p.')
      PAUSE
C       Definitions:
C
C       Variable
C       Name            Description
C       ----            -----------
C
C       A               Number of particles added
C       APOR            Actual aquifer porosity as a decimal
C       ANORM           Random number (-6,6) with normal
C                       distribution
```

```
C     AT          Aquitard thickness (ft)
C     BOT         Aquifer base elevation (ft)
C     CH          Aquifer top elevation (ft)
C     CIW         Injection well contaminant concentration
C                 (mg/L)
C     CPB         Pit or basin contaminant concentration
C                 (mg/L)
C     CSCC        Sourcebed contaminant concentration (mg/L)
C     CONC        Contaminant concentration at
C                 each node (mg/L)
C     CONSOR      Pre-existing contaminant concentration
C                 at each node (mg/L)
C     CONSNK      Contaminant concentration of water
C                 flowing
C                 into sink (mg/L)
C     D           Days moved for a particle
C     DD          Distance particle has moved (ft)
C     DE          Random portion of delta (days)
C     DELTA       Time increment particle are allowed to
C                 move (days)
C     DEL         Time sub-increment particles are
C                 are allowed to move (days)
C     DELX        Grid spacing in X-direction (ft)
C     DELY        Grid spacing in Y-direction (ft)
C     DISPL       Aquifer longitudinal dispersivity (ft)
C     DISPT       Transverse dispersivity (ft)
C     DK          Aquifer distribution coefficient (mL/g)
C     DKL         Aquitard distribution coefficient (mL/g)
C     DX          Spatial increment added to X-coordinate
C                 over which particles will be placed or moved
C     DY          Spatial increment added to Y-coordinate
C                 over which particles will be placed or moved
C     EPOR        Aquifer effective porosity as a decimal
C     EPORA       Aquitard effective porosity as a decimal
C     GLP         Length of pit or basin in ft
C     H           Aquifer head elevation (ft)
C     H1          Sourcebed head elevation (ft)
C     H2          Aquifer layer 2 head elevation (ft)
C     HL          Half-life of radioactive isotope (days)
C     HPB         Aquifer head beneath pit or basin (ft)
C     I           Index variable
C     IIJ         I-coordinate of injection well
C     ISS         I-coordinate of sink
C     IP          I-coordinate of production well sink
C     IVG1,IVG2   I-coordinates of grid block corners
C     JIJ         J-coordinate of injection well
C     JSS         J-coordinate of sink
C     JVG1,JVG2   J-coordinates of grid block corners
C     JP          J-coordinate of production well sink
C     J           Index variable
C     K           Index variable
C     L           Index variable
```

```
C   MARK      Table of sink coordinates, the value stored
C             in mark is the sink number, if mark is 0,
C             there is no sink
C   MAXP      Maximum number of particles in system
C   MOVE      Number of sink that captures a particle
C   NGB       Number of grid blocks
C   NSORS     Number of slug sources
C   NIJ       Number of injection wells
C   NPITS     Number of contaminant pits or basins
C   NSSN      Number of stream sink nodes
C   NSA       Number of rectangular source areas
C   NPCS      Number of particles captured by sink
C   NC        Number of columns
C   NP        Number of particles in system
C   NPART     Number of particles at each node
C   NPUMP     Number of production well sinks
C   NR        Number of rows
C   PA        Aquitard vertical hydraulic conductivity
C             (gpd/sq ft)
C   PIT       Pit or basin vertical hydraulic
C             conductivity (gpd/sq ft)
C   PTH       Pit or basin bed thickness (ft)
C   PLEAK     Pit or basin leakance (gpd/ft)
C   PSE       Elevation of water surface in pit
C             or basin in ft
C   PS        Streambed vertical hydraulic
C             conductivity (gpd/sq ft)
C   PST       Pit or basin water surface elevation (ft)
C   P1        Aquifer hydraulic conductivity in Y-direction
C             between I,J and I,J+1 (gpd/sq ft)
C   P2        Aquifer hydraulic conductivity in X-direction
C             between I,J and I+1,J (gpd/sq ft)
C   P3        Aquitard vertical hydraulic conductivity
C             (gpd/sq ft)
C   P4        Aquifer layer 1 vertical hydraulic
C             conductivity (gpd/sq ft)
C   P5        Aquifer layer 2 vertical hydraulic
C             conductivity (gpd/sq ft)
C   PL        Total amount of contaminant (pounds)
C   PM        Particle mass (pounds)
C   Q         Production well sink discharge (gpd)
C   RHO       Bulk mass density of aquifer skeleton (g/cu
C             cm)
C   RHOL      Bulk mass density of aquitard skeleton
C             (g/cu m)
C   RD1       Aquifer retardation factor
C   RD1V      Aquitard retardation factor
C   R,RS      Simulation factor (gpd/ft)
C   RATE      Injection well recharge rate (gpd)
C   RD        Elevation of streambed base (ft)
C   RH        Elevation of stream surface or
C             sourcebed head (ft)
```

```
C     SEED1      Seed for random number generation
C     STRW       Streambed width in ft
C     STRL       Streambed length in ft
C     THS        Streambed thickness (ft)
C     THIC       Aquifer thickness beneath pit or basin (ft)
C     TABLE      Mass of particles in sink (pounds)
C     THICKC     Aquitard thickness (ft)
C     THICK3     Aquitard thickness (ft)
C     THICK4     Aquifer layer 1 thickness (ft)
C     THICK5     Aquifer layer 2 thickness (ft)
C     THS        Streambed thickness (ft)
C     TMAP       Time since particles were first allowed
C                to move (days)
C     V          Array containing particle velocities (ft/day)
C     VX         Velocity in X-direction (ft/day)
C     VY         Velocity in Y-direction (ft/day)
C     VV         Array containing particle velocities (ft/day)
C     WP         Width of pit or basin in ft
C     X1         I-coordinate of upper left corner of
C                rectangular source area
C     X2         I-coordinate of lower right corner of
C     X3         I-coordinate of end of line source
C     X4         I-coordinate of center of circle source area
C     X4C        I-coordinate of point on a line along the
C                X-axis,through center of circle source area,
C                and on the circumference of the circle
C                source area
C     X8         I-coordinate of beginning of line source
C     X7         I-coordinate of point source
C                rectangular source area
C     Y1         J-coordinate of upper left corner of
C                rectangular source area
C     Y2         Y-coordinate of lower right corner of
C                rectangular source area
C     Y3         Y-coordinate of end of line source
C     Y4         Y-coordinate of center of circle source area
C     Y7         Y-coordinate of point source
C     Y8         Y-coordinate of beginning of line source
      END
```

```
C
      SUBROUTINE INIT
C
C     IN THIS SUBROUTINE THE MARK ARRAY
C     TMAP AND NP ARE ZEROED OUT AND ANC
C     AND ANR ARE INITIALIZED
C
C     METACOMMANDS
C
$LARGE
C
C     DECLARATIONS
C
      COMMON/NSUB1/NC,NR
      COMMON/NSUB7/NP,NPART(50,50)
      COMMON/NSUB6/MAXP,NSORS,NIJ,NSSN,MARK(50,50),NRAD
      COMMON/AREA8/TMAP,ANC,ANR,TABLE(124)
C
      TMAP=0.
      NP=0
      ANC=NC+.5
      ANR=NR+.5
      DO 20 J=1,NR
      DO 10 I=1,NC
      MARK(I,J)=0
   10 CONTINUE
   20 CONTINUE
      END
```

```
C
      SUBROUTINE INJIN(MFL)
C
C     THIS SUBROUTINE GENERATES PARTICLES TO
C     SIMULATE INJECTION WELL CONTAMINATION.
C     INJECTION WELL GEOMETRY AND STRENGTH ARE DEFINED
C     AND PARTICLES ARE DISTRIBUTED AT POINT RANDOMLY
C     IN TIME
C
C     METACOMMANDS
C
$LARGE
C
C     DECLARATIONS
C
      COMMON/NSUB6/MAXP,NSORS,NIJ,NSSN,MARK(50,50),NRAD
      COMMON/AREA6/DELTA,THICKC(50,50),PS(50,50)
      COMMON/NSUB7/NP,NPART(50,50)
      COMMON/AREA7/PM,DISPL,DISPT,EPOR,APOR
      COMMON/NSUB3/IJJ(24),JIJ(24)
      COMMON/AREA5/X(5001),Y(5001),V(50,50,2)
      COMMON/AREA8/TMAP,ANC,ANR,TABLE(124)
      COMMON/NSUB1/NC,NR
      COMMON/AREA3/RD1,HL,PLEAK(50),Q(50,50),RD1V
      COMMON/AREA2/DELX(51),DELY(51),P1(50,50),P2(50,50)
C
      OPEN(5,FILE='LPT1:')
      DO 100 K=1,NIJ
      WRITE(*,30)K
   30 FORMAT('0','Injection well number=',I3)
      WRITE(5,40)K
   40 FORMAT('0','Injection well number=',I3)
      WRITE(*,50)
   50 FORMAT(1X,'Enter real number injection well water'/1X,
     *'concentration in mg/l  '\)
      READ(*,*)CIW
      WRITE(5,60)CIW
   60 FORMAT(1X,'Injection well water concentration'/1X,
     *'in mg/l=',F10.3)
      WRITE(*,70)
   70 FORMAT(1X,'Enter real number injection rate'/1X,
     *'in gpm  '\)
      READ(*,*)RATE
      WRITE(5,80)RATE
   80 FORMAT(1X,'Injection rate in gpm=',F9.2)
      RATE=RATE*1440
```

```
 82 PL=RATE*DELTA*CIW*8.3453/1E6
    IF(MFL.EQ.1)WRITE(*,84)
 84 FORMAT('0','COMPUTATIONS ARE IN PROGRESS')
    MFL=2
    P=PL
 90 IF(RAN(0).GT.P/PM)GOTO 100
    P=P-PM
    XX=IJJ(K)
    YY=JIJ(K)
    DE=DELTA*RAN(0)
    IF(MOVE(XX,YY,DE).EQ.0)CALL ADD(XX,YY)
    GOTO 90
100 CONTINUE
    CLOSE (5)
    END
```

```
C
      SUBROUTINE LOCSOR(MFL,NN8)
C
C     THIS SUBROUTINE GENERATES PARTICLES TO
C     SIMULATE LOCAL PIT OR BASIN CONTAMINANT
C     LEAKAGE WITH KNOWN SOURCE CONCENTRATION.
C     PIT OR BASIN GEOMETRY AND LEAKANCE ARE DEFINED
C     AND PARTICLES ARE DISTRIBUTED RANDOMLY WITHIN
C     PIT OR BASIN AREA AND IN TIME
C
C     METACOMMANDS
C
$LARGE
C
C     DECLARATIONS
C
      COMMON/NSUB2/NPUMP,NPITS
      COMMON/NSUB7/NP,NPART(50,50)
      COMMON/NSUB6/MAXP,NSORS,NIJ,NSSN,MARK(50,50),NRAD
      COMMON/AREA7/PM,DISPL,DISPT,EPOR,APOR
      COMMON/AREA6/DELTA,THICKC(50,50),PS(50,50)
      COMMON/AREA3/RD1,HL,PLEAK(50),Q(50,50),RD1V
      COMMON/AREA5/X(5001),Y(5001),V(50,50,2)
      COMMON/AREA8/TMAP,ANC,ANR,TABLE(124)
      COMMON/NSUB1/NC,NR
      COMMON/AREA2/DELX(51),DELY(51),P1(50,50),P2(50,50)
      COMMON/NSUB4/IP(24),JP(24),ISS(100),JSS(100)
      COMMON/NSUB8/NAL,NLSK,NTT,NPE,NN1,IWS,NLC,NI1
      COMMON/AREA1/RH(50,50),R(50,50),RD(50,50),RS(50,50)
      COMMON/ASUB2/CH(50,50),BOT(50,50),H(50,50)
C
      OPEN(5,FILE='LPT1:')
      IF(NN8.EQ.2)THEN
      DO 14 K=1,NSSN
      WRITE(*,2)K
    2 FORMAT('0','Stream source node number=',I3)
      WRITE(5,4)K
    4 FORMAT('0','Stream source node number=',I3)
      WRITE(*,6)
    6 FORMAT(1X,'Enter real number stream source contaminant'/1X,
     *'concentration during simulation period in mg/l  '\)
      READ(*,*)CSTR
      WRITE(5,8)CSTR
    8 FORMAT(1X,'Stream source contaminant concentration'/1X,
     *'during simualtion period in mg/l=',F10.3)
      IF(MFL.EQ.1)WRITE(*,9)
```

```
 9 FORMAT('0','COMPUTATIONS ARE IN PROGRESS')
   MFL=2
   I=ISS(K)
   J=JSS(K)
   RATE=(RH(I,J)-H(I,J))*RS(I,J)
   IF(H(I,J).LT.RD(I,J))RATE =(RH(I,J)-RD(I,J))*RS(I,J)
   PL=RATE*CSTR*8.3453*DELTA/1E6
10 P=PL
12 IF(RAN(0).GT.P/PM)GOTO 14
   P=P-PM
   XX=ISS(K)
   YY=JSS(K)
   DE=DELTA*RAN(0)
   IF(MOVE(XX,YY,DE).EQ.0)CALL ADD(XX,YY)
   GOTO 12
14 CONTINUE
   ENDIF
   IF(NLC.EQ.1)THEN
   DO 270 K=1,NPITS
   WRITE(*,120)K
120 FORMAT('0','Pit or basin source number=',I3)
   WRITE(5,130)K
130 FORMAT('0','Pit or basin source number=',I3)
   WRITE(*,140)
140 FORMAT(1X,'Enter real number pit or basin source '/1X,
  *'contaminant concentration during simulation time'/1X,
  *'increment in mg/l  '\)
   READ(*,*)CPB
   WRITE(5,150)CPB
150 FORMAT(1X,'Pit or basin source contaminant concentration'/1X,
  *'in mg/l=',F10.3)
   WRITE(*,160)
160 FORMAT(1X,'Enter real number I and J coordinates'/1X,
  *'of pit or basin source area upper left corner separated'/1X,
  *'a comma  '\)
   READ(*,*)X1,Y1
   WRITE(5,170)X1,Y1
170 FORMAT(1X,'I and J coordinates of pit or basin'/1X,
  *'source area upper left corner=',F5.2,',',F5.2)
   WRITE(*,180)
180 FORMAT(1X,'Enter real number I and J coordinates'/1X,
  *'of pit or basin source area lower right corner'/1X,
  *'separated by a comma  '\)
   READ(*,*)X2,Y2
   WRITE(5,190)X2,Y2
190 FORMAT(1X,'Pit or basin source area lower right corner'/1X,
  *'I and J coordinates=',F5.2)
   DX=ABS(X1-X2)
   DY=ABS(Y1-Y2)
   WRITE(*,200)
200 FORMAT(1X,'Enter real number average elevation of'/1X,
  *'water surface in pit or basin source area in ft  '\)
```

```
      READ(*,*)PSE
      WRITE(5,210)PSE
210 FORMAT(1X,'Average elevation of water surface in pit'/1X,
   *'or basin source area in ft=',F9.2)
      WRITE(*,220)
220 FORMAT(1X,'Enter real number average head in aquifer'/1X,
   *'during simulation time increment under pit or basin'/1X,
   *'source area in ft  '\)
      READ(*,*)HPB
      WRITE(5,230)HPB
230 FORMAT(1X,'Average head in aquifer during simulation time'/1X,
   *'increment at pit or basin source area in ft=',F9.2)
      WRITE(*,232)
232 FORMAT(1X,'Enter real number width of pit or basin'/1X,
   *'in ft  '\)
      READ(*,*)WP
      WRITE(5,233)WP
233 FORMAT(1X,'Width of pit or basin in ft=',F9.2)
      WRITE(*,234)
234 FORMAT(1X,'Enter real number length of pit or basin'/1X,
   *'in ft  '\)
      READ(*,*)GLP
      WRITE(5,236)GLP
236 FORMAT(1X,'Length of pit or basin in ft=',F9.2)
      RATE=(PSE-HPB)*PLEAK(K)*WP*GLP
      PL=RATE*DELTA*CPB*8.3453/1E6
      IF(MFL.EQ.1)WRITE(*,245)
245 FORMAT('0','COMPUTATIONS ARE IN PROGRESS')
      MFL=2
252 P=PL
260 IF(RAN(0).GT.P/PM)GOTO 270
      P=P-PM
      XX=X1+DX*RAN(0)
      YY=Y1+DY*RAN(0)
      DE=DELTA*RAN(0)
      IF(MOVE(XX,YY,DE).EQ.0)CALL ADD(XX,YY)
      GOTO 260
270 CONTINUE
      ENDIF
272 CLOSE (5)
      END
```

```
C
      SUBROUTINE MAP
C
C     THIS SUBROUTINE PROVIDES A HARD COPY
C     OF NODAL DISTRIBUTION OF PARTICLES
C
C     METACOMMANDS
C
$LARGE
C
C     DECLARATIONS
C
      COMMON/NSUB1/NC,NR
      COMMON/NSUB7/NP,NPART(50,50)
      COMMON/AREA7/PM,DISPL,DISPT,EPOR,APOR
      COMMON/AREA8/TMAP,ANC,ANR,TABLE(124)
C
      OPEN(5,FILE='LPT1:')
      WRITE(5,10)
   10 FORMAT('0','NO. OF PARTICLES RESIDING IN MODEL GRIDS:')
      WRITE(5,12)PM
   12 FORMAT('0','Particle mass in pounds=',F10.3)
      WRITE(5,14)NP
   14 FORMAT(1X,'Number of particles within grid=',I5)
      WRITE(5,15)
   15 FORMAT(1X,'If number of particles within grid plus'/1X,
     *'number of particles captured by sinks is less than'/1X,
     *'number of injected particles then some particles have'/1X,
     *'positions outside grid')
      WRITE(5,16)TMAP
   16 FORMAT(1X,'Time in days=',F9.3)
      N1=1
      N2=15
      IF(NC.LT.N2)N2=NC
   17 WRITE(5,18)
   18 FORMAT('0','ROW                              COLUMN')
   20 WRITE(5,22) (I,I=N1,N2)
   22 FORMAT(3X,15I5)
      DO 26 J=1,NR
      WRITE(5,24)J,(NPART(I,J),I=N1,N2)
   24 FORMAT(I3,15I5)
   26 CONTINUE
      IF(N2.EQ.NC)GO TO 40
      N1=N1+15
      N2=N2+15
      IF(NC.LT.N2)N2=NC
      GO TO 17
      CLOSE (5)
   40 END
```

```
C
      FUNCTION MOVE(XX,YY,DEL)
C
C     THIS FUNCTION MOVES PARTICLES, KEEPS TAB ON
C     ON SINKS, LIMITS MOVEMENT OF PARTICLES AND
C     COLLECTS PARTICLE DATA
C
C     METACOMMANDS
C
$LARGE
C
C     DECLARATIONS
C
      COMMON/NSUB1/NC,NR
      COMMON/NSUB7/NP,NPT(50,50)
      COMMON/NSUB6/MAXP,NSORS,NIJ,NSSN,MARK(50,50),NRAD
      COMMON/AREA3/RD1,HL,PLEAK(50),Q(50,50),RD1V
      COMMON/AREA7/PM,DISPL,DISPT,EPOR,APOR
      COMMON/AREA8/TMAP,ANC,ANR,TABLE(124)
      COMMON/AREA2/DELX(51),DELY(51),P1(50,50),P2(50,50)
      COMMON/AREA5/X(5001),Y(5001),V(50,50,2)
C
      D=DEL
    1 IF(D.LE.0.0)GO TO 4
      CALL VELO(XX,YY,VX,VY)
C
C     LIMIT DISTANCE PARTICLE CAN MOVE
C
      IXX=XX
      IYY=YY
      T=AMIN1(D,.2*ABS(DELX(IXX)/VX),.2*ABS(DELY(IYY)/VY))
      IF(VX.EQ.0.)T=AMIN1(D,.2*ABS(DELY(IYY)/VY))
      IF(VY.EQ.0.)T=AMIN1(D,.2*ABS(DELX(IXX)/VX))
      IF(VX.EQ.0. .AND. VY.EQ.0.)T=D
      D=D-T
      DX=VX*T
      DY=VY*T
      SIGNX=1.0
      SIGNY=1.0
      IF(DX.LT.0.)SIGNX=-1.0
      IF(DY.LT.0.)SIGNY=-1.0
      IF(DX.EQ.0.)DX=1E-20
      IF(DY.EQ.0.)DY=1E-20
      PHI=ATAN(DY/DX)
      DD=SQRT(DX**2+DY**2)/RD1
      IF(DD.EQ.0.)DD=1E-20
```

```
      RL=(SQRT(2*DISPL*DD)/DD)*ANORM(0)
      RT=(SQRT(2*DISPT*DD)/DD)*ANORM(0)
      IXX=MIN0(NC,INT(XX+.5))
      IYY=MIN0(NR,INT(YY+.5))
      DX=DD*ABS(COS(PHI))*SIGNX
      DY=DD*ABS(SIN(PHI))*SIGNY
      XX=XX+(DX+RL*DX+RT*DY)/(DELX(IXX))
      YY=YY+(DY+RL*DY-RT*DX)/(DELY(IYY))
C
C      LIMIT PARTICLE MOVEMENT TO FIELD
C
      XX=AMIN1(ANC,AMAX1(XX,1.0))
      YY=AMIN1(ANR,AMAX1(YY,1.0))
C
C      SUM SINK PARTICLE MASS INTO TABLE
C
      IX=MIN0(NC,INT(XX+.5))
      IY=MIN0(NR,INT(YY+.5))
      MOVE=MARK(IX,IY)
      IF(MOVE.EQ.0)GO TO 1
      TABLE(MOVE)=TABLE(MOVE)+PM
      RETURN
C
C      TABULATE LOCATION OF PARTICLES
C
    4 IX=XX+.5
      IY=YY+.5
      IF(IX.GT.NC)GOTO 10
      IF(IY.GT.NR)GOTO 10
      NPT(IX,IY)=NPT(IX,IY)+1
   10 MOVE=0
      END
```

```
C
      SUBROUTINE PRESOR
C
C     THIS SUBROUTINE GENERATES PARTICLES TO
C     SIMULATE PRE-EXISTING CONTAMINANT CONDITIONS.
C     NODAL CONTAMINANT LOAD IS DEFINED AND PARTICLES
C     ARE DISTRIBUTED RANDOMLY IN AREA CENTERED AT NODES
C     AS A SLUG
C
C     METACOMMANDS
C
$LARGE
C
C     DECLARATIONS
C
      COMMON/NSUB1/NC,NR
      COMMON/ASUB1/CONSNK(50,50),CONC(50,50),
     *CSCC(50,50),CONSOR(50,50)
      COMMON/AREA2/DELX(51),DELY(51),P1(50,50),P2(50,50)
      COMMON/ASUB2/CH(50,50),BOT(50,50),H(50,50)
      COMMON/AREA7/PM,DISPL,DISPT,EPOR,APOR
      COMMON/AREA5/X(5001),Y(5001),V(50,50,2)
      COMMON/NSUB6/MAXP,NSORS,NIJ,NSSN,MARK(50,50),NRAD
      COMMON/NSUB7/NP,NPART(50,50)
      COMMON/AREA8/TMAP,ANC,ANR,TABLE(124)
C
      DO 30 J=1,NR
      DO 20 I=1,NC
      PL=CONSOR(I,J)*DELX(I)*DELY(J)*(CH(I,J)-BOT(I,J))*
     *APOR*62.4/1E6
      P=PL
   10 IF(RAN(0).GT.P/PM)GOTO 20
      P=P-PM
      XX=((I-1)+.5)+1.0*RAN(0)
      YY=((J-1)+.5)+1.0*RAN(0)
      CALL ADD(XX,YY)
      GOTO 10
   20 CONTINUE
   30 CONTINUE
      END
```

```
C
      SUBROUTINE PROFIL
C
C     METACOMMANDS
C
$LARGE
C
C     THIS SUBROUTINE CONVERTS PLAN VIEW AQUIFER
C     UNIT OR LAYER CONTAMINANT CONCENTRATION
C     DATA BASE INTO PROFILE DATA BASE DISPLAY
C
C     DECLARATIONS
C
      CHARACTER*14FNAME
      DIMENSION CNODE(50,10)
C
C     DEFINE OUTPUT DEVISE
C
      OPEN(5,FILE='LPT1:')
C
   30 WRITE(*,40)
   40 FORMAT('0','Enter 1 to convert plan view aquifer'/1X,
     *'unit or layer contaminant concentration data base'/1X,
     *'into profile data base display or 2 for no display  '\)
      READ(*,*)MP1
      IF(MP1.NE.1.AND.MP1.NE.2)GOTO 30
      IF(MP1.EQ.2)GOTO 810
      WRITE(*,100)
  100 FORMAT('0','Enter integer number of aquifer units or'/1X,
     *'layers on profile (must be < 11)  '\)
      READ(*,*)NUL
      WRITE(5,102)NUL
  102 FORMAT('0','Number of aquifer units or layers on'/1X,
     *'profile=',I3)
      WRITE(*,130)
  130 FORMAT(1X,'Enter integer number of nodes on'/1X,
     *'profile (must be < 51)  '\)
      READ(*,*)NODE
      WRITE(5,140)NODE
  140 FORMAT(1X,'Number of nodes on profile=',I3)
      DO 392 L=1,NODE
      WRITE(*,300)L
  300 FORMAT('0','Node number=',I3)
      DO 390 N=1,NUL
      WRITE(*,380)N
  380 FORMAT(1X,'Aquifer unit or layer number=',I3)
```

```
      WRITE(*,382)
382 FORMAT(1X,'Enter real number contaminant concentration'/1X,
   *'at node in aquifer unit or layer in mg/l  '\)
      READ(*,*)CNODE(L,N)
390 CONTINUE
392 CONTINUE
460 WRITE(*,470)
470 FORMAT('0','Enter 1 for screen display of head profile'/1X,
   *'or 2 to continue program  '\)
      READ(*,*)N1C
      IF(N1C.NE.1.AND.N1C.NE.2)GOTO 460
      IF(N1C.EQ.1)THEN
      WRITE(*,472)
472 FORMAT('0','Use Ctrl-Num Lock keys to'/1X,
   *'control screen scrolling')
      PAUSE
      WRITE(*,480)
480 FORMAT('0','CONCENTRATION IN AQUIFER UNITS OR'/1X,
   *'AQUIFER'/1X,
   *'LAYERS IN MG/L:')
500 WRITE(*,510)
510 FORMAT('0','UNIT OR                    NODE NUMBER'/1X,
   *'LAYER NO.'/)
      N1=1
      N2=8
      IF(NODE.LT.N2)N2=NODE
610 WRITE(*,620)(I,I=N1,N2)
620 FORMAT(3X,8I9)
      DO 640 J=1,NUL
      WRITE(*,622)J,(CNODE(I,J),I=N1,N2)
622 FORMAT(I3,8F9.3)
640 CONTINUE
      IF(N2.EQ.NODE)GOTO 650
      N1=N1+8
      N2=N2+8
      IF(NODE.LT.N2)N2=NODE
      GOTO 610
650 ENDIF
660 WRITE(*,665)
665 FORMAT('0','Enter 1 for hard copy of concentration'/1X,
   *'profile or 2 to continue program  '\)
      READ(*,*)N2C
      IF(N2C.NE.1.AND.N2C.NE.2)GOTO 660
      IF(N2C.EQ.1)THEN
      N1=1
      N2=8
      IF(NODE.LT.N2)N2=NODE
      WRITE(5,680)
680 FORMAT('0','CONCENTRATION IN AQUIFER UNITS OR'/1X,
   *'AQUIFER'/1X,
   *'LAYERS IN MG/L:')
690 WRITE(5,700)
```

```
700 FORMAT('0','UNIT OR                              NODE NUMBER'/1X,
   *'LAYER NO.'/)
710 WRITE(5,720)(I,I=N1,N2)
720 FORMAT(3X,8I9)
    DO 740 J=1,NUL
    WRITE(5,734)J,(CNODE(I,J),I=N1,N2)
734 FORMAT(I3,8F9.3)
740 CONTINUE
    IF(N2.EQ.NODE)GOTO 750
    N1=N1+8
    N2=N2+8
    IF(NODE.LT.N2)N2=NODE
    GOTO 710
750 ENDIF
    CLOSE (5)
810 END
```

```
C
      FUNCTION RAN(R)
C
C     THIS FUNCTION GENERATES RANDOM NUMBERS
C     BETWEEN 0 AND 1
C
C
C
      DATA ISEED/0/
      I1=29
      I2=217
      I3=1024
      ISEED=MOD(ISEED*I1+I2,I3)
      RAN=REAL(ISEED)/I3
      END
```

```
C
      SUBROUTINE SLUGIN(CSLUG)
C
C     THIS SUBROUTINE GENERATES PARTICLES TO
C     SIMULATE SLUG SOURCES. CIRCLE,LINE,RECTANGLE,
C     OR POINT SOURCE GEOMETRY AND STRENGTH AREA ARE
C     DEFINED
C     AND PARTICLES ARE DISTRIBUTED UNIFORMLY IN SPACE
C
C     METACOMMANDS
C
$LARGE
C
C     DELARATIONS
C
      DIMENSION CSLUG(50,50)
      COMMON/NSUB6/MAXP,NSORS,NIJ,NSSN,MARK(50,50),NRAD
      COMMON/AREA5/X(5001),Y(5001),V(50,50,2)
      COMMON/AREA7/PM,DISPL,DISPT,EPOR,APOR
      COMMON/NSUB7/NP,NPART(50,50)
      COMMON/AREA8/TMAP,ANC,ANR,TABLE(124)
      COMMON/AREA2/DELX(51),DELY(51),P1(50,50),P2(50,50)
      COMMON/ASUB2/CH(50,50),BOT(50,50),H(50,50)
C
      OPEN(5,FILE='LPT1:')
      N1=0
      N2=0
      N3=0
      N4=0
      WRITE(*,5)
    5 FORMAT('0','Slug injection of contaminants through wells'/1X,
     *'or landfill or surface water beds may be simulated')
      DO 320 K=1,NSORS
      WRITE(*,6)K
    6 FORMAT('0','Slug source number=',I4)
      WRITE(5,7)K
    7 FORMAT('0','Slug source number=',I4)
      WRITE(*,8)
    8 FORMAT(1X,'Enter real number total contaminant'/1X,
     *'slug load in pounds  '\)
      READ(*,*)PL
      WRITE(5,9)PL
    9 FORMAT(1X,'Total contaminant slug load in pounds=',F10.3)
      PT1=PL/PM
      M=ANINT(PT1)
   10 WRITE(*,20)
```

```
 20 FORMAT('0','Circle means injecting particles uniformly'/1X,
   *'around the circumference of a circle, line means'/1X,
   *'injecting particles uniformly along a line, rectangle'/1X,
   *'means injecting particles uniformly within a rectang-'/1X,
   *'ular area, and point means injecting particles at a'/1X,
   *'point. Enter 1 for circle, 2 for line, 3 for rectangle,'/1X,
   *'or 4 for point  '\)
    READ(*,*)NSL
    IF(NSL.NE.1.AND.NSL.NE.2.AND.NSL.NE.3.AND.NSL.NE.4)GOTO 10
    IF(NSL.EQ.1)THEN
    N11=N1+1
    WRITE(*,50)N11
 50 FORMAT('0','Circle source number=',I3)
    WRITE(5,60)N11
 60 FORMAT('0','Circle source number=',I3)
    WRITE(*,70)
 70 FORMAT(1X,'Enter real number circle source center'/1X,
   *'coordinates I and J separated by a comma  '\)
    READ(*,*)X4,Y4
    WRITE(5,80)X4,Y4
 80 FORMAT(1X,'Circle center coordinates I,J=',F5.2,',',F5.2)
    WRITE(*,90)
 90 FORMAT(1X,'Enter real number I coordinate of a point'/1X,
   *'on a line along x-axis,through center of circle source'/1X,
   *'area,and on the circumference of the circle  '\)
    READ(*,*)X4C
    WRITE(5,100)X4C
100 FORMAT(1X,'I coordinate of a point on a line along the'/1X,
   *'x-axis,through center of circle source area,and on'/1X,
   *'the circumference of the circle=',F5.2)
    CR=ABS(X4C-X4)
    AN=0.0
    AG=2*3.14159/M
    DO 110 L=1,M
    TH=AG*L+AN
    XA=X4+CR*SIN(TH)
    YA=Y4+CR*COS(TH)
    XX=XA
    YY=YA
    I=NINT(XX)
    J=NINT(YY)
    CPMSLUG=PM*1E6/(DELX(I)*DELY(J)*(CH(I,J)-BOT(I,J))*
   *APOR*62.4)
    CSLUG(I,J)=CSLUG(I,J)+CPMSLUG
    CALL ADD(XX,YY)
110 CONTINUE
    N11=N1+1
    ENDIF
    IF(NSL.EQ.2)THEN
    N22=N2+1
    WRITE(*,120)N22
120 FORMAT('0','Line source number=',I3)
```

```
      WRITE(5,130)N22
  130 FORMAT('0','Line source number=',I3)
      WRITE(*,140)
  140 FORMAT(1X,'Enter real number beginning I and J'/1X,
     *'coordinates of line source separated by comma  '\)
      READ(*,*)X8,Y8
      WRITE(5,150)X8,Y8
  150 FORMAT(1X,'Beginning coordinates of line source'/1X,
     *'I,J=',F5.2,',',F5.2)
      WRITE(*,160)
  160 FORMAT(1X,'Enter real number end I and J coordinates'/1X,
     *'of line source separated by a comma  '\)
      READ(*,*)X3,Y3
      WRITE(5,170)X3,Y3
  170 FORMAT(1X,'End coordinates of line source'/1X,
     *'I,J=',F5.2,',',F5.2)
      XX=X8
      YY=Y8
      I=NINT(XX)
      J=NINT(YY)
      CPMSLUG=PM*1E6/(DELX(I)*DELY(J)*(CH(I,J)-BOT(I,J))*
     *APOR*62.4)
      CSLUG(I,J)=CSLUG(I,J)+CPMSLUG
      CALL ADD(XX,YY)
      IF(M.GT.1)XX=X3
      IF(M.GT.1)YY=Y3
      I=NINT(XX)
      J=NINT(YY)
      CPMSLUG=PM*1E6/(DELX(I)*DELY(J)*(CH(I,J)-BOT(I,J))*
     *APOR*62.4)
      CSLUG(I,J)=CSLUG(I,J)+CPMSLUG
      IF(M.GT.1)CALL ADD(XX,YY)
      IF(M.EQ.2)GOTO 190
      DO 180 L=3,M
      XX=X8+(L-2)*(X3-X8)/(M-1)
      YY=Y8+(L-2)*(Y3-Y8)/(M-1)
      I=NINT(XX)
      J=NINT(YY)
      CPMSLUG=PM*1E6/(DELX(I)*DELY(J)*(CH(I,J)-BOT(I,J))*
     *APOR*62.4)
      CSLUG(I,J)=CSLUG(I,J)+CPMSLUG
      CALL ADD(XX,YY)
  180 CONTINUE
      N22=N2+1
  190 ENDIF
      IF(NSL.EQ.3)THEN
      N33=N3+1
      WRITE(*,200)N33
  200 FORMAT('0','Rectangle source number=',I3)
      WRITE(5,210)N33
  210 FORMAT('0','Rectangle source number=',I3)
      WRITE(*,220)
```

```
220 FORMAT(1X,'Enter real number rectangle source area upper'/1X,
   *'left corner coordinates I and J separated by'/1X,
   *'comma  '\)
    READ(*,*)X5,Y5
    WRITE(5,230)X5,Y5
230 FORMAT(1X,'Rectangle source area upper left corner'/1X,
   *'coordinates I,J=',F5.2,',',F5.2)
    WRITE(*,240)
240 FORMAT(1X,'Enter real number rectangle source area lower'/1X,
   *'right corner coordinates I and J separated by'/1X,
   *'comma  '\)
    READ(*,*)X6,Y6
    WRITE(5,250)X6,Y6
250 FORMAT(1X,'Rectangle source area lower right corner'/1X,
   *'coordinates I,J=',F5.2,',',F5.2)
    XO=X5
    YO=Y5
    DX1=(X6-X5)/M**.5/2
    DY1=(Y6-Y5)/M**.5/2
    Y1=YO-DY1
    DO 260 L=1,M
    X1=XO+DX1
    IF(X1.GT.X6)XO=X5
    IF(X1.GT.X6)Y1=Y1-2*DY1
    IF(X1.GT.X6)X1=X5+DX1
    IF(X1.LT.X5)X1=X5+((X6-X5)/2)
    IF(Y1.LT.Y5)Y1=Y5+((Y6-Y5)/2)
    IF(Y1.GT.Y6)Y1=Y5+((Y6-Y5)/2)
    XO=XO+2*DX1
    XX=X1
    YY=Y1
    I=NINT(XX)
    J=NINT(YY)
    CPMSLUG=PM*1E6/(DELX(I)*DELY(J)*(CH(I,J)-BOT(I,J))*
   *APOR*62.4)
    CSLUG(I,J)=CSLUG(I,J)+CPMSLUG
    CALL ADD(XX,YY)
260 CONTINUE
    N33=N3+1
    ENDIF
    IF(NSL.EQ.4)THEN
    N44=N4+1
    WRITE(*,270)N44
270 FORMAT('0','Point source number=',I3)
    WRITE(5,280)N44
280 FORMAT('0','Point source number=',I3)
    WRITE(*,290)
290 FORMAT(1X,'Enter real number point source coordinates'/1X,
   *'I and J separated by comma  '\)
    READ(*,*)X7,Y7
    WRITE(5,300)X7,Y7
300 FORMAT(1X,'Point source coordinates I,J=',F5.2,',',F5.2)
```

```
      DO 310 L=1,M
      XX=X7
      YY=Y7
      I=NINT(XX)
      J=NINT(YY)
      CPMSLUG=PM*1E6/(DELX(I)*DELY(J)*(CH(I,J)-BOT(I,J))*
     *APOR*62.4)
      CSLUG(I,J)=CSLUG(I,J)+CPMSLUG
      CALL ADD(XX,YY)
  310 CONTINUE
      N44=N4+1
      ENDIF
  320 CONTINUE
      CLOSE (5)
      END
```

```
C
      SUBROUTINE SNKCON(NSOD)
C
C      THIS SUBROUTINE PRINTS CONTAMINANT CONCENTRATION
C      OF SINK DISCHARGE
C
C      METACOMMANDS
C
$LARGE
C
C      DECLARATIONS
C
      COMMON/ASUB1/CONSNK(50,50),CONC(50,50),
     *CSCC(50,50),CONSOR(50,50)
      COMMON/NSUB1/NC,NR
      COMMON/ASUB2/CH(50,50),BOT(50,50),H(50,50)
      COMMON/NSUB6/MAXP,NSORS,NIJ,NSSN,MARK(50,50),NRAD
      COMMON/AREA3/RD1,HL,PLEAK(50),Q(50,50),RD1V
      COMMON/AREA1/RH(50,50),R(50,50),RD(50,50),RS(50,50)
      COMMON/AREA8/TMAP,ANC,ANR,TABLE(124)
      COMMON/AREA6/DELTA,THICKC(50,50),PS(50,50)
C
      OPEN(5,FILE='LPT1:')
      DO 20 J=1,NR
      DO 10 I=1,NC
      CONSNK(I,J)=0.0
   10 CONTINUE
   20 CONTINUE
      DO 70 I=1,NC
      DO 60 J=1,NR
      IF(MARK(I,J).LT.1)GO TO 60
      MOVE=MARK(I,J)
      IF(TABLE(MOVE).EQ.0)GO TO 60
      IF(Q(I,J).GT.0.0)GOTO 30
      IF(RS(I,J).EQ.0.0)GOTO 60
      IF((H(I,J)-RH(I,J)).EQ.0.0)GOTO 60
      CONSNK(I,J)=TABLE(MOVE)*1E6/((H(I,J)-RH(I,J))*
     *DELTA*RS(I,J)*8.3453)
      IF(H(I,J).LT.RD(I,J))CONSNK(I,J)=TABLE(MOVE)*
     *1E6/((H(I,J)-RD(I,J))*DELTA*RS(I,J)*8.3453)
      IF(NSOD.EQ.1)CONSNK(I,J)=CONSNK(I,J)/RD1
      IF(NSOD.EQ.2)CONSNK(I,J)=CONSNK(I,J)*RD1
      GOTO 50
   30 CONSNK(I,J)=TABLE(MOVE)*1E6
     */((Q(I,J)*DELTA*8.3453))
      IF(NSOD.EQ.1)CONSNK(I,J)=CONSNK(I,J)/RD1
```

```
   IF(NSOD.EQ.2)CONSNK(I,J)=CONSNK(I,J)*RD1
50 IF(TABLE(MOVE).GT.0)WRITE(*,52)I,J,CONSNK(I,J)
52 FORMAT('0','Contaminant concentration of production'/1X,
  *'well sink discharge or aquifer discharge to stream'/1X,
  *'at node with I,J coordinates',I3,',',I3,' in mg/l=',/1X,
  *F10.3)
   IF(TABLE(MOVE).GT.0)WRITE(5,55)I,J,CONSNK(I,J)
55 FORMAT('0','Contaminant concentration of production'/1X,
  *'well sink discharge or aquifer discharge to stream'/1X,
  *'at node with I,J coordinates ',I3,',',I3,' in mg/l=',/1X,
  *F10.3)
60 CONTINUE
70 CONTINUE
   CLOSE (5)
   END
```

```
C
      SUBROUTINE SUMMRY
C
C     THIS SUBROUTINE PRINTS THE NUMBER OF PARTICLES
C     CAPTURED BY SINK DURING TIME INCREMENT DELTA
C
C     METACOMMANDS
C
$LARGE
C
C     DECLARATIONS
C
      COMMON/NSUB1/NC,NR
      COMMON/NSUB6/MAXP,NSORS,NIJ,NSSN,MARK(50,50),NRAD
      COMMON/AREA7/PM,DISPL,DISPT,EPOR,APOR
      COMMON/AREA8/TMAP,ANC,ANR,TABLE(124)
      COMMON/AREA3/RD1,HL,PLEAK(50),Q(50,50),RD1V
C
      OPEN(5,FILE='LPT1:')
      DO 60 J=1,NR
      DO 50 I=1,NC
      IF(MARK(I,J).LT.1)GOTO 50
      MOVE=MARK(I,J)
      NPCS=TABLE(MOVE)/PM
      IF(NRAD.EQ.1)NPCS=NPCS*(.5)**(TMAP/HL)
      IF(TABLE(MOVE).GT.0)WRITE(*,40)I,J,NPCS
   40 FORMAT('0','Number of particles captured by sink'/1X,
     *'at node with I,J coordinates',I3,',',I3,' =',I5)
      IF(TABLE(MOVE).GT.0)WRITE(5,42)I,J,NPCS
   42 FORMAT('0','Number of particles captured by sink'/1X,
     *'at node with I,J coordinates',I3,',',I3,' =',I5)
   50 CONTINUE
   60 CONTINUE
      CLOSE (5)
      END
```

```
C
      SUBROUTINE VELO(XX,YY,VX,VY)
C
C     THIS SUBROUTINE CALCULATES PARTICLE VELOCITIES
C     KEEPS PARTICLES INSIDE MODEL AND INSURES PROPER
C     VELOCITIES
C
C     METACOMMANDS
C
$LARGE
C
C     DECLARATIONS
C
      COMMON/NSUB1/NC,NR
      COMMON/AREA5/X(5001),Y(5001),V(50,50,2)
C
      LIJ(NR,J,YA)=MIN0(NR-1,MAX0(1,J+IFIX(SIGN(1.0,YA-0.5))))
      I=XX
      IF(I.LE.0)I=1
      IF(I.GE.NC)I=NC-1
      J=YY
      IF(J.LE.0)J=1
      IF(J.GE.NR)J=NR-1
      XA=XX-I
      YA=YY-J
      VX=YA*V(I,J+1,2)+(1.-YA)*V(I,J,2)
      VY=XA*V(I+1,J,1)+(1.-XA)*V(I,J,1)
      JY=LIJ(NR,J,YA)
      IX=LIJ(NC,I,XA)
      IF(ABS(VY).LE.1E-20)GO TO 10
      IF(JY.EQ.J)GO TO 10
      YZ=ABS(YA-0.5)
      VY=VY*(1.0-YZ)+YZ*(XA*V(I+1,JY,1)+(1.0-XA)*V(I,JY,1))
   10 IF(IX.EQ.I)GO TO 20
      IF(ABS(VX).LE.1E-20)GO TO 20
      XZ=ABS(XA-0.5)
      VX=VX*(1.0-XZ)+XZ*(YA*V(IX,J+1,2)+(1.0-YA)*V(IX,J,2))
   20 CONTINUE
      IF(ABS(VX).LE.1E-20)VX=1E-20
      IF(ABS(VY).LE.1E-20)VY=1E-20
      END
```

```
C
      FUNCTION V(I,J,K)
C
C     THIS FUNCTION PROVIDES VELOCITIES TO SUBROUTINE VELO
C
C     METACOMMANDS
C
$LARGE
C
C     DECLARATIONS
C
      COMMON/AREA5/X(5001),Y(5001),VV(50,50,2)
C
      V=VV(I,J,K)
      END
```

```
C
      SUBROUTINE XYZDTC(CONC)
C
C     THIS SUBROUTINE CONVERTS CALCULATED CONCENTRATION
C     ARRAY INTO XYZ DATA FOR USE IN GRAPHICS PROGRAMS
C
C     METACOMMANDS
C
$LARGE
C
C     DECLARATIONS
C
      CHARACTER*14FNAME
      DIMENSION XSUM(50),YSUM(50),CONC(50,50)
      COMMON/NSUB1/NC,NR
      COMMON/AREA2/DELX(51),DELY(51),P1(50,50),P2(50,50)
C
   10 WRITE(*,20)
   20 FORMAT('0','Enter 1 to convert calculated concentration'/1X,
     *'array data to a standard ASCII file containing'/1X,
     *'one X-coordinate (X),Y-coordinate (Y),concentration-'/1X,
     *'coordinate (Z) entry per line; the (X,Y,Z) values on a'/1X,
     *'line are separated by a comma. Enter 2 for no'/1X,
     *'conversion file  '\)
      READ(*,*)NZ
      IF(NZ.NE.1.AND.NZ.NE.2)GOTO 10
   21 IF(NZ.EQ.1)THEN
      DO 30 J=1,NR
      IF(J.EQ.1)THEN
      YSUM(J)=0.5*DELY(J)
      GOTO 30
      ENDIF
      YSUM(J)=YSUM(J-1)+0.5*DELY(J-1)+0.5*DELY(J)
   30 CONTINUE
      DO 40 I=1,NC
      IF(I.EQ.1)THEN
      XSUM(I)=0.5*DELX(I)
      GOTO 40
      ENDIF
      XSUM(I)=XSUM(I-1)+0.5*DELX(I-1)+0.5*DELX(I)
   40 CONTINUE
      WRITE(*,70)
   70 FORMAT('0','Enter a valid X,Y,Z data base specification'/1X,
     *'filename with extension DAT of ASCII file to be exported'/1X,
     *'such as B:FILE6.DAT in apostrophes  '\)
      READ(*,*)FNAME
```

```
      OPEN(6,FILE=FNAME,STATUS='NEW')
      DO 100 J=1,NR
      DO 90 I=1,NC
      WRITE(6,80)XSUM(I),YSUM(J),CONC(I,J)
 80   FORMAT(1X,F12.2,',',F12.2,',',F9.3)
 90   CONTINUE
100   CONTINUE
      CLOSE (6)
      ENDIF
      END
```

Appendix C
Diskette Instructions

Instructions for using the three diskettes included with this book are given in this appendix. It is assumed that the reader is familiar with the operation of the computer and its peripherals and understands the Disk Operating System manual supplied with the computer.

These instructions were written for an IBM Personal Computer or "compatible" running PC-DOS or MS-DOS Version 2.0 or later (PS-DOS is a trademark of International Business Machines, Inc., and MS-DOS is a trademark of Microsoft Corp.) with 640 K RAM available memory, a monitor, and a PC-DOS or MS-DOS "compatible" printer. The programs can take advantage of, but do not require, an 8087 numeric coprocessor. Instructions assume you have two diskette drives; if you have one diskette drive or a fixed drive, consult your Disk Operating System manual. This program requires 640K of RAM for its operation. Printers, other devices, and resident programs using RAM other than DOS are not supported. Their use will cause an out-of-memory error. A printer is required.

The diskettes do not contain DOS commands because of copyright restrictions prohibiting the distribution of bootable diskettes. Insert your DOS diskette in drive A, insert diskette 1 in drive B, and turn on the power switch. Respond to prompts by entering the date and time. In response to the system prompt A> request a directory of diskette 1 using the DOS command: dir b:. The following files should be displayed (though not necessarily in this order): DATPR.FOR, DRAIN.FOR, EVAP.FOR, FILEI.FOR, FILEO.FOR, FWEL.FOR, GWFL3D.FOR, INFO1.FOR, INFO2.FOR, MINE.FOR, PARPEN.FOR, PROFIL.FOR, RESLT.FOR, RIVER.FOR, SIMEQ.FOR, WATBL.FOR, WELSTOR.FOR, and XYZDT.FOR. Repeat this process with the other two diskettes. Diskette 2 should contain the file GWFL3D.EXE (248216 bytes of memory) and diskette 3 should contain the following files: ADD.FOR, ADVAN.FOR, ANORM.FOR, AREA3D-.FOR, ARESNK.FOR, ARESOR.FOR, CLEAR.FOR, CONMAP.FOR, CONSTR.FOR, CONTIN.FOR, DATPRN.FOR, DISPER.FOR, FILEI1.FOR, FILEI2.FOR, FILEI3.FOR, FILEO1.FOR, FILEO2.FOR, FILEO3-.FOR, GWTR3D.FOR, HEAD.FOR, INFOM.FOR, INIT-.FOR, INJIN.FOR, LOCSOR.FOR, MAP.FOR, MOVE.-FOR, PRESOR.FOR, PROFIL.FOR, RAN.FOR, SUMMRY.FOR, SNKCON.FOR, SLUGIN.FOR, VELO.FOR, VIJK.FOR, XYZDTC.FOR, and GWTR3D.EXE (176728 bytes of memory).

With your DOS diskette in drive A, remove diskette 3 from drive B and insert a blank diskette in drive B. In response to the system prompt A> , format the blank diskette using the DOS command: format b:/s. Copy the file GWFL3D.EXE on the formatted diskette by removing your DOS diskette from drive A, inserting diskette 2 in drive A, and in response to the system prompt A> using the DOS command copy GWFL3D.EXE b:. The directory of formatted working diskette A is now complete. Repeat this process to create formatted working diskette B containing the file GWTR3D.EXE. The executable files were created with the Microsoft FORTRAN Compiler Version 3.2 using the default compiler and library options so that the programs will run on any environment (with or without 8087) and give the most accurate results possible.

To use the working diskettes in drive A, in response to prompts enter the date and time and then in response to the system prompt A> enter either GWFL3D or

GWTR3D depending on whether you desire to simulate groundwater flow or contaminant transport. The specified program will be loaded and run. The user must turn on the printer or else an error will occur. The user then responds to data base and other prompt/option statements. To exit before program end, press simultaneously Ctrl-Break keys. Preprocessors are integrated with other parts of the programs. Execution times for most data bases will range from a few minutes to an hour. Large data bases may result in execution times of several hours. Serious and unexpected data base input errors or other user errors may result in program lock, and pressing simultaneously Ctrl and Break keys or other keys will not unlock the program. In this case, remove the working diskette from drive A, turn off the computer, and start over again.

The author has taken due care in preparing the source codes and creating the diskettes, including research and testing to ensure their accuracy and effectiveness. Neither the author nor the publisher make any warranty of any kind, expressed or implied, with regard to the performance of the source codes or diskettes. No warranties, expressed or implied, are made by the author or the publisher that the source codes and diskettes are free of error, or are consistent with any particular standard of merchantability, or that they meet the reader's requirements for any particular application. In no event shall the author or the publisher be liable for incidental or consequential damages in connection with or arising from the furnishing, performance, or use of the source codes and diskettes. Support for the source codes diskettes and diskettes is this book and is not otherwise available via telephone or mail.

Although this book and its programs are copyrighted, the reader is authorized to make one machine-readable copy of each program for personal use. Distribution of the machine-readable programs (either as copied by the reader or supplied with this publication) is not authorized. Before use, the programs should be verified with respect to known problem solutions over the range of available data. Programs will give meaningful results only for reasonable problems. The programs represent basic and general models that may have to be modified by the user for efficient application to specific field problems. Although the programs will produce reliable calculations for a wide variety of field problems, the user is cautioned that in some cases the accuracy and efficiency of the models can be affected significantly by the selection of values for certain user-defined options. For example, the accuracy and precision of contaminant transport solutions are sensitive to the number of particles placed in each grid block. In some cases, particle mass and the time increment may have to be significantly decreased to achieve reasonable accuracy and definition in time-concentration curves.

If this is your first experience with numerical groundwater modeling, read Marsily (1986) before running programs. If you have not read Prickett and Lonnquist (1971) and Prickett, et al. (1981), do so before running programs. The reader should read the contents of this book, including the preface, before running the programs. If this is your first experience with FORTRAN 77, read Fuori, et al. (1986) before attempting to modify the source codes. Program operating information is presented in Chapters 2 and 4 and displayed on the screen (see information statements in source codes in Appendixes A and B). The source codes are provided so that the user may modify programs as required. The user should solve the simple problems described in Appendix E before attempting to solve other problems. Unreasonable data base entry and/or prompt/option selection may result in program failure. Several days of program operation may be required before the user is thoroughly familiar with the programs. The diskettes included with this book should not be thought of as commercial software, but rather as aids for learning about numerical groundwater modeling. Particular applications are the responsibility of the user, not the author or the publisher.

Appendix D
Model Capabilities and Limitations

GWFL3D and GWTR3D model capabilities and limitations are summarized in this appendix to help prevent potential program misuse and frustration due to exaggerated program expectations. In general, models are idealizations of reality and blindly accept imperfect data bases. The user is responsible for making meaningful conceptual generalizations and approximations and knowing how far these can be pushed before errors become so great as to invalidate any program results. It should be recognized that data bases are generally incomplete and subject to varying degrees of uncertainty. Some insight into the uncertainty of results should be gained by the user through sensitivity studies. Contaminant migration problems involve a level of complexity and related uncertainty that is an order of magnitude greater than for flow problems. A flow model which is supposed to serve as a basis for contaminant migration calculation must satisfy a higher standard of accuracy than a flow model which is used for water balance calculations. Flow feature selection can have a profound effect on contaminant migration calculations.

The finite-difference model for program GWFL3D simulates one-, two-, and quasi-three-dimensional groundwater flow with constant density and viscosity through saturated uniformly porous aquifer systems. Water movement through the unsaturated zone, flow in the saturated zone with variable density, and flow in fractured rock aquifer systems are not simulated. Homogeneous and heterogeneous horizontal and vertical aquifer system hydraulic characteristics and irregular aquifer system boundaries and thicknesses; up to four aquifer units (aquifer-aquitard), two or four adjoining aquifer layers with inter-aquifer flow transfer, or two adjoining aquifer units and two aquifer layers; infinite aquifer systems and barrier, constant head (recharge), specified head, and specified flow boundaries; and nonleaky artesian, leaky artesian, conversion from artesian to water table, and water table conditions are simulated.

The head in the sourcebed of the uppermost aquifer unit must be constant and water table conditions must prevail in the uppermost aquifer layer. It is assumed that an aquiclude underlies the lowermost aquifer or aquifer layer. Aquitard storativity may be simulated in which case it is assumed that transient leaky effects from previous pumping periods have dissipated. Delayed gravity yield under water table conditions simulation requires that only one aquifer unit be specified and there can be no flowing wells, drains, mines, evapotranspiration, or areal recharge. Induced streambed infiltration can be modeled only in the uppermost aquifer unit or layer. Evapotranspiration cannot be modeled at streambed nodes. Time-variable uniform areal recharge and evapotranspiration can be modeled only in the uppermost aquifer unit or layer; and time- and space-variable streambed location, width, length, thickness, and vertical hydraulic conductivity can be simulated.

The number of time increments in the simulation period must be < 51 except if delayed gravity yield is simulated when the number of time increments must be < 21. Uniform (square) or variable grid spacings may be specified, however, grid spacings must be the same in all aquifer units or layers. The number of grid columns or rows is limited to 20. High precision head and flow calculations depend on the proper selection of the number and duration of time increments and the magnitude of grid spacings.

The number of production or injection (recharge) wells

must be $<$ 51 and wells must be located at grid nodes in any aquifer or layer. The number of observation wells for which time-head tables are to be prepared must be $<$ 25 and the wells must be located at grid nodes in any aquifer or layer. Time-variable production well discharge and injection well recharge rates are simulated. The head in the production well is calculated on the basis of the average head in the production well grid block. Partial penetration impacts are calculated for wells only in aquifer units and not aquifer layers. Only one production well may have well storage capacity, In addition, well storage simulation assumes that discharge derived from the aquifer or layer varies linearly and the well discharge rate is constant during the simulation period. The number of flowing wells is limited to 50 and the flowing wells must be located at grid nodes in any artesian aquifer or layer. It is assumed that water levels do not decline below the top of flowing well casings. All flowing wells are capped or opened simultaneously. Only one multi-unit or layer fully penetrating production or injection well located at a grid node is simulated. The well radius must be the same in all aquifer units or layers open to the well. Flow with head differences from one aquifer unit or layer to another when the well is not operating is simulated.

Drains must be located in the uppermost aquifer unit or layer at grid nodes. Evapotranspiration cannot be modeled at drain nodes. It is assumed that the head in the drain is at the elevation of the top of the drain and water levels do not decline below the top of the drain. Time- and space-variable drain location, top elevation, backfill thickness and hydraulic conductivity, diameter, and length are simulated. Mine nodes may be located in any aquifer unit or layer at least two nodes from boundaries or other discharge or recharge facilities. Evapotranspiration cannot be modeled at mine nodes. Mine nodes may be active and advancing, inactive, or backfilled. Fully or partially penetrating mines are simulated. Mine shafts are assumed to be sealed and mine drifts are assumed to be centered at mine nodes. The mine drift cavity is assumed to extend to distances midway between adjacent grid nodes. The grid spacing must be uniform (square) at and near mine nodes. Time-variable mine floor elevation, mine penetration, and backfill hydraulic characteristics are simulated.

The discrete parcel random walk model for program GWTR3D simulates one-, two-, or quasi-three-dimensional contaminant migration with constant density, viscosity, and dispersion coefficients; sorption; and radioactive decay in saturated uniform porous aquifer systems. Contaminant concentration is assumed not to affect the liquid density and viscosity. It is assumed that sorption reactions are fast and reversible, the isotherm is linear, and there is only one contaminant in solution and no interaction of contaminants. Contaminant migration through the unsaturated zone, contaminant migration in the saturated zone with variable density and viscosity, contaminant migration in fractured rock aquifer systems, oxidation/reduction reactions, ion exchange, complexation, co-solvation, immiscible phase partitioning, microbial population dynamics, substrata utilization, biotransformation, adoptation, and cometabolism are not simulated.

Horizontal contaminant migration from fully penetrating slug and/or continuous point sources towards sinks in one aquifer or aquifer layer during one time increment with or without pre-existing concentration conditions, homogeneous and heterogeneous aquifer system hydraulic characteristics, and irregular aquifer system boundaries and thicknesses is simulated. Aquifer or aquifer layer actual and effective porosities are assumed to be uniform in space. Vertical contaminant migration is simulated with multiple program runs through areal sink and source calculations. For example, during program run 1, vertical leakage of contaminants from aquifer 1 through an aquitard and into aquifer 2 may be simulated as areal sink mass in aquifer 1. During program run 2, this areal sink mass from aquifer 1 becomes areal source mass in aquifer 2. Contaminant migration during multiple time increments is simulated with multiple program runs. Any uniform or irregular pre-existing contaminant loads are distributed at the start of the simulation time increment randomly in spaces centered at grid nodes and extending midway to surrounding nodes. Calculation precision is dependent on the proper choice of the number of particles (must be $<$ 5001) representing contaminants, particle mass, simulation time increment, and grid spacing.

Contaminant migration analysis requires a pre-flow modeling effort with fine enough resolution to define the velocity field with sufficient detail to represent the advective movement of contaminants. Required sourcebed, aquifer, and aquifer layer head data bases may be imported from a file or user defined from the keyboard. Uniform (square) or variable grid spacings may be specified. The number of grid columns or rows must be $<$ 51 and must match the number of columns and rows used in creating any imported head data base file. If preexisting contaminant conditions are simulated, the model grid and any preexisting contaminant loads must extend beyond the area under investigation to allow migration of pre-existing contaminants to and from that area.

Continuous sources such as contaminant injection wells, landfills, pits, basins, ditches, surface water bodies, and areal leakage are simulated. Injection well contaminant sources must be located at grid nodes. The number of injection wells must be $<$ 51. The injection well liquid concentration and injection rate are assumed to be constant. Up to 50 contaminant sources consisting of leachate leakage through a landfill, pit, basin, ditch, or surface water bed with known contaminant concentration or known contaminant load into an aquifer or aquifer layer may be located at grid nodes or within grid blocks. The thickness and hydraulic conductivity of beds are assumed to be uniform in space. Contaminant concentration must be constant. The average elevation of the water surface in landfills, pits, basins, ditches, or surface water bodies and the average head in the aquifer or aquifer layer, during the simulation period are used in calculations. Up to 50 contaminant sources rectangular in area with known contaminant load, unknown solute concentration, and contaminant load injected directly into the aquifer or aquifer layer are simulated. These sources may be located at grid nodes or within grid blocks. The contaminant load is distributed randomly throughout the rectangular area. Areal contaminant sources consisting of continuous leakage through an aquitard into the aquifer or aquifer layer assuming the aquifer head is above the aquitard base and sourcebed contaminant mass is not depleted may be simulated unless areal leakage sinks are present. The sourcebed head, aquitard hydraulic characteristics, and sourcebed contaminant concentration may be variable in space. Areal contaminant loads are distributed randomly in spaces centered at grid nodes and extending midway to surrounding nodes. The simulation time increment for the areal source must be the same as that used in areal sink calculations.

Slug sources with contaminants uniformly distributed at the start of the simulation period around the circumference of a circle, along a line, within a rectangular, or at a point are simulated. Slug sources may be located at grid nodes or within grid blocks.

Surface water sinks located at up to 100 grid nodes are simulated. The surface water bed width, length, vertical hydraulic conductivity, base elevation, and thickness and water surface elevation are uniform in space. Up to 25 production well sinks may be simulated at grid nodes. The production well discharges are assumed to be constant. Areal leakage sinks located at grid nodes and consisting of vertical contaminant migration from one aquifer layer to an adjacent aquifer layer or vertical contaminant migration from one aquifer through an aquitard and into another aquifer with contaminant mass depletion in the aquifer or layer is simulated. With areal leakage sinks, sourcebed or aquifer layer heads, aquitard and/or aquifer layer vertical hydraulic conductivity and thickness may vary in space. It is assumed that the average concentration in the aquifer or layer during the simulation time increment is the sum of the concentrations at start and end of the simulation time increment divided by two. Heads must be average heads during the simulation time increment. There can be sorption in the aquifer, aquifer layer, or aquitard. It is assumed that vertical dispersion is negligible and fairly uniform vertical head losses occur within the area under investigation. Vertical head losses must reflect average head conditions during the travel time through the aquifer layer or aquitard. Heads in the aquifer or aquifer layer receiving the contaminant load from area sinks must reflect average head conditions during a simulation time increment after the travel time through the aquifer layer or aquitard.

A large amount of printer paper is required when the option to obtain a hard copy of the time-constant data base in GWFL3D or GWTR3D is chosen. Horizontal aquifer hydraulic conductivity and transmissivity values adjusted for variable grids (see pages 16–18) are printed as part of the time-constant data base hard copy. Actual values are printed only when a square grid is specified. If the time-constant data base is imported and the wrong drive is specified in the data base filename, then a fatal error will occur. Mine width must be equal to mine length, and the square grid spacing at a mine node must be equal to the mine width or length. Aquifer system and facility features are often too complex to be simulated in intimate detail, and the modeler must exercise considerable ingenuity in approximating features. Data base preparation is a time- and effort-consuming task usually unappreciated by modelers.

Appendix E
Example Program Input/Output Displays

Two example input/output displays to guide the user through initial GWFL3D and GWTR3D program operation are presented in this Appendix. Data bases constitute input and computation results constitute output. Most of the information presented was displayed on the screen and on paper by the printer.

Example E.1 GWFL3D Input/Output Display

The Example E.1 model consists of a simple uniformly porous leaky artesian aquifer with uniform aquifer horizontal hydraulic conductivity of 189 gpd/sq ft, aquifer storativity of 0.0002, aquifer specific yield of 0.2, aquifer thickness of 8 ft, aquitard vertical hydraulic conductivity of 0.11 gpd/sq ft, and aquitard thickness of 14 ft. The aquifer system has no boundaries or discontinuities. There is no aquifer layering or aquitard storativity, leaky artesian conditions prevail in the uppermost aquifer unit, and artesian conditions are simulated in the uppermost aquifer unit. The initial head in the aquifer is 0.0 ft as is the constant head in the sourcebed above the aquitard. The aquifer top elevation is −20 ft and the aquifer base elevation is −28 ft. There is one fully penetrating production well with I, J coordinates of 5, 5; a constant discharge rate of 25 gpm; and an effective radius of .25 ft. There is one time-drawdown observation well with I, J coordinates of 6, 5. The uniform model

10×10 grid has a spacing of 500 ft. Drawdown is calculated at three times: 0.01 day, 0.1 day, and 1.0 day.

```
MODEL DATA BASE SPECIFICATION:

Number of time increments=  3
Time increment number=  1
Time at end of time increment in days=       .01000
Time increment in days=       .01000
Time increment number=  2
Time at end of time increment in days=       .10000
Time increment in days=       .09000
Time increment number=  3
Time at end of time increment in days=      1.00000
Time increment in days=       .90000
Number of grid columns= 10
Number of grid rows= 10
Number of units or layers=  1

Constant head in sourcebed above
uppermost unit in ft=      .00
Number of production wells=  1
Number of injection wells=  0
Number of time-head observation wells=  1

Production well number  1
I-coordinate=  5  J-coordinate=  5
Unit or layer number= 1

Observation well number  1
I-coordinate=  6  J-coordinate=  5
unit or layer number= 1

Production well number=  1
Production well discharge rate in gpm=    25.00

Head or drawdown in production well 1
 in ft=   -23.44   (at .01 day)
```

COMPUTATION NODAL RESULTS:

TIME(DAYS)= 1.00000

HEAD OR DRAWDOWN (FT) IN UNIT OR LAYER 1:

ROW	COLUMN 1	2	3	4	5	6	7	8
1	.00	-.01	-.01	-.02	-.03	-.02	-.01	-.01
2	-.01	-.01	-.03	-.05	-.08	-.05	-.03	-.01
3	-.01	-.03	-.07	-.17	-.29	-.17	-.07	-.02
4	-.02	-.05	-.17	-.50	-1.21	-.50	-.16	-.05
5	-.03	-.08	-.29	-1.21	-5.31	-1.21	-.29	-.07
6	-.02	-.05	-.17	-.50	-1.21	-.50	-.16	-.05
7	-.01	-.03	-.07	-.16	-.29	-.16	-.07	-.02
8	-.01	-.01	-.02	-.05	-.07	-.05	-.02	-.01
9	.00	.00	-.01	-.02	-.02	-.02	-.01	.00
10	.00	.00	.00	-.01	-.01	-.01	.00	.00

ROW	9	10
1	.00	.00
2	.00	.00
3	-.01	.00
4	-.02	-.01
5	-.02	-.01
6	-.02	-.01
7	-.01	.00
8	.00	.00
9	.00	.00
10	.00	.00

Observation Well Number 1

Time (Days)	Head (Ft)
.01000	-.03
.10000	-.48
1.00000	-1.21

Example E.2 GWTR3D Input/Output Display

The Example E.2 model consists of a simple uniformly porous water table aquifer with uniform horizontal hydraulic conductivity of 100 gpd/sq ft, thickness of 20 ft, actual porosity of 0.32, effective porosity of 0.23, longitudinal dispersivity of 15 ft, and transverse dispersivity of 3 ft. The aquifer top elevation is 702 ft and the aquifer base elevation is 682 ft. The aquifer has no boundaries or discontinuities. There is one fully penetrating continuous rectangular source area with a known contaminant load of 50 pounds; upper right corner I, J coordinates of 2.0, 5.0; and lower left corner I, J coordinates of 2.2, 5.3. The particle mass is 0.5 pounds so that 100 particles are injected into the aquifer. There is no initial pre-existing contaminant concentration, slug contaminant source, radioactive decay, injection well, contaminant source load from areal leakage sink, source due to contaminant leakage through an aquitard, local contaminant source leakage through a pit bed, or sink. A uniform gradient head data base is created by the user at the keyboard with a seepage velocity of 1 ft/day and a water table elevation of 702 ft along the leftmost grid column. Importation of data files is not required. Contaminant migration is calculated for a time period of 200 days. The uniform model 10×10 grid has a spacing of 50 ft.

```
MODEL DATA BASE SPECIFICATION:

Number of grid columns= 10
Number of grid rows= 10
Default (square) grid spacing in ft=   50.
Grid setting is uniform (square)
Duration of time increment in days=  200.000
Default aquifer or aquifer layer hydraulic
conductivity in gpd/sq ft=   100.00
Default top of aquifer or aquifer layer elevation
in ft=   702.00
Default base of aquifer or aquifer layer
elevation in ft=   682.00
```

```
Actual aquifer or aquifer layer porosity
as a decimal=  .3200
Effective aquifer or aquifer layer porosity
as a decimal=  .2300
Particle mass in pounds=      .500
Maximum number of particles in model= 5000
Longitudinal dispersivity in ft=  15.000
Transverse dispersivity in ft=   3.000


Number of rectangular areas=  1

Rectangular source area number=  1

Total contaminant load during this
time increment in pounds=    50.000
I,J coordinates of upper left corner
of rectangular source area= 2.00, 5.00
I,J coordinates of lower right corner
of rectangular source area= 2.20, 5.30


TIME (DAYS)=  200.000

CONCENTRATION VALUES (MG/L):
```

ROW	COLUMN 1	2	3	4	5	6	7	8
1	.000	.000	.000	.000	.000	.000	.000	.000
2	.000	.000	.000	.000	.000	.000	.000	.000
3	.000	.000	.000	.000	.000	.000	.000	.000
4	.000	.000	.501	.501	3.005	4.006	.000	.000
5	2.003	8.013	13.021	4.507	3.005	.000	1.002	2.504
6	.000	.000	.000	1.502	4.006	2.504	.000	.000
7	.000	.000	.000	.000	.000	.000	.000	.000
8	.000	.000	.000	.000	.000	.000	.000	.000
9	.000	.000	.000	.000	.000	.000	.000	.000
10	.000	.000	.000	.000	.000	.000	.000	.000

ROW	COLUMN	
	9	10
1	.000	.000
2	.000	.000
3	.000	.000
4	.000	.000
5	.000	.000
6	.000	.000
7	.000	.000
8	.000	.000
9	.000	.000
10	.000	.000

NO. OF PARTICLES RESIDING IN MODEL GRIDS:

Particle mass in pounds= .500
Number of particles within grid= 100
If number of particles within grid plus
number of particles captured by sinks is less than
number of injected particles then some particles have
positions outside grid
Time in days= 200.000

ROW	COLUMN									
	1	2	3	4	5	6	7	8	9	10
1	0	0	0	0	0	0	0	0	0	0
2	0	0	0	0	0	0	0	0	0	0
3	0	0	0	0	0	0	0	0	0	0
4	0	0	1	1	6	8	0	0	0	0
5	4	16	26	9	6	0	2	5	0	0
6	0	0	0	3	8	5	0	0	0	0
7	0	0	0	0	0	0	0	0	0	0
8	0	0	0	0	0	0	0	0	0	0
9	0	0	0	0	0	0	0	0	0	0
10	0	0	0	0	0	0	0	0	0	0

References

Ahlstrom, S.W., H.P. Foote, R.C. Arnett, C.R. Cole, and R.J. Serne. 1977. Multicomponent Mass Transport Model Theory and Numerical Implementation (Discrete Parcel Random Walk Version). Battelle Report BNWL 2127 for Energy and Research Development Administration.

Bear, J. 1972. Dynamics of Fluids in Porous Media. Elsevier Science Publishing Co., Inc., New York.

Bear, J., and A. Verruijt. 1987. Modeling Groundwater Flow and Pollution. D. Reidel Publishing Company, Norwell, MA.

Beljin, M.S. 1987. Representation of Individual Wells in Two-Dimensional Ground Water Modeling. In Proceedings of the NWWA Conference on Solving Ground Water Problems with Models. National Water Well Association, Dublin, OH.

Bennett, G.D., A.L. Kontis, and S.P. Larson. 1982. Representation of Multiaquifer Well Effects in Three-Dimensional Ground-Water Flow Simulation. Ground Water. Vol. 20, No. 3.

Bourke, P.D. 1987. A Contouring Subroutine. BYTE Publications Inc., New York, June.

Bredehoeft, J.D., and G.F. Pinder. 1970. Digital Analysis of Area Flow in Multiaquifer Groundwater Systems: A Quasi Three-Dimensional Model. Water Resources Research. Vol. 6, No. 3.

Crain, I.K. 1970. Computer Interpolation and Contouring

of Two-Dimensional Data: A Review. Geoexploration. Vol. 8.

Driscoll, F.G. 1986. Groundwater and Wells. Second Edition. Johnson Division. St. Paul, MN.

Etter, D.M. 1987. Structured FORTRAN 77 for Engineers and Scientists. Second Edition. Benjamin-Cummings Publishing Company, Inc., Menlo Park, CA.

Ferris, J.G., D.B. Knowles, R.H. Brown, and R.W. Stallman. 1962. Theory of Aquifer Tests. U.S. Geological Survey. Water Supply Paper 1536-E.

Fowler, J. 1984. IBM PC/XT Graphics Book. Prentice-Hall, Inc., Englewood Cliffs, NJ.

Franke, O.L., T.E. Reilly, and G.D. Bennett. 1985. Definition of Boundary and Initial Conditions in the Analysis of Saturated Ground-Water Flow Systems-An Introduction. U.S. Geological Survey Open-File Report 84–458.

Fuori, W.M., S. Gaughran, L. Gioia, and M. Fuori. 1986. FORTRAN 77 Elements of Programming Style. Hayden Book Company. Hasbrouck Heights, NJ.

Hantush, M.S. 1964. Hydraulics of Wells. In Advances in Hydroscience. Vol. 1. Van Te Chow, Ed. Academic Press, Inc., San Diego, CA.

Hearn, D., and M.P. Baker. 1983. Computer Graphics for the IBM Personal Computer. Prentice-Hall, Inc., Englewood Cliffs, NJ.

Huyakorn, P.S., and Pinder, G.F. 1983. Computational Methods in Subsurface Flow. Academic Press, Inc., San Diego, CA.

Jacob, C.E. 1947. Drawdown Test to Determine Effective Radius of Artesian Well. Transactions of the American Society of Civil Engineers, Vol. 112.

Karplus, W.J. 1958. Analog Simulation. McGraw-Hill Book Company, New York.

Kinzelbach, W. 1987. Methods for the Simulation of Pollutant Transport in Ground Water-A Model Comparison. In Proceedings of NWWA Conference on Solving Ground Water Problems with Models. National Water Well Association, Dublin, OH.

Knuth, A. 1969. The Art of Computer Programming. Vol. 2, Chap. 3. Addison-Wesley Publishing Co., Reading, MA.

Konikow, L.F., and J.D. Bredehoeft. 1978. Computer Model of Two-Dimensional Solute Transport and Dispersion in Ground Water. U.S Geological Survey. Techniques of Water-Resources Investigations. Book 7, Chapter C2.

Korites, B.J. 1982. Data Plotting Software for Micros. Kern Publications, Duxbury, MA.

Lancaster, P., and K. Salkauskas. 1986. Curve and Surface Fitting. Academic Press, Inc., San Diego, CA.

Marsily, Ghislain de. 1986. Quantitative Hydrogeology. Academic Press, Inc., San Diego, CA.

McDonald, M.G., and A.W. Harbaugh. 1984. A Modular Three-Dimensional Finite-Difference Ground-Water Flow Model. U.S. Geological Survey. Open-File Report 83-875.

McLain, D.H. 1974. Drawing Contours from Arbitrary Data Points. Computing Journal. Vol. 17.

Neuman, S.P. 1979. Perspective on "Delayed Yield." Water Resources Research. Vol. 15, No. 4.

OSM. 1981. Ground Water Model Handbook. U.S. Office of Surface Mining. Denver, CO. H-D3004-021-81-1062D.

Peaceman, D.W. 1983. Interpretation of Well-Block Pressures in Numerical Reservoir Simulation with Nonsquare Grid Blocks and Anisotropic Permeability. Society of Petroleum Engineers Journal, Vol. 23, No. 3.

Pedrosa, O.A., and K. Aziz. 1986. Use of a Hybrid Grid in Reservoir Simulation. Society of Petroleum Engineers. Reservoir Engineering. Vol. 1, No. 6.

Pelto, C.R., T.A. Elkins, and H.A. Boy. 1968. Automatic Contouring of Irregularly Spaced Data. Geophysics. Vol. 33.

Prickett, T.A., and C.G. Lonnquist. 1971. Selected Digital Computer Techniques for Groundwater Resource Evaluation. Illinois State Water Survey. Bulletin 55.

Prickett, T.A., T.S. Naymik, and C.G. Lonnquist. 1981. A

"Random Walk" Solute Transport Model for Selected Groundwater Quality Evaluations. Illinois State Water Survey. Bulletin 65.

Raudkivi, A.J. and R.A. Callander. 1976. Analysis of Groundwater Flow. John Wiley & Sons, New York.

Rushton, K.R., and S.C. Redshaw. 1979. Seepage and Groundwater Flow. John Wiley & Sons, New York.

Simons, S.L. November, 1983. Make Fast and Simple Contour Plots on a Microcomputer. BYTE Publications Inc., New York.

Streltsova, T.D. 1972. Unsteady Radial Flow in an Unconfined Aquifer. Water Resources Research Vol. 8, No. 4.

Todd, D.K. 1980. Groundwater Hydrology. Second Edition. John Wiley & Sons, New York.

Trescott, P.C., G.F. Pinder, and S.P. Larson. 1976. Finite-Difference Model for Aquifer Simulation in Two Dimensions with Results of Numerical Experiments. U.S Geological Survey. Techniques of Water-Resources Investigations. Chapter C1, Book 7.

Uffink, G.J.M. 1985. A Random Walk Method for the Simulation of Macrodispersion in a Stratified Aquifer. In Relation of Groundwater Quantity and Quality. International Association of Hydrological Sciences Publication No. 146.

Voisinet, D.D. 1986. Introduction to Computer-Aided Drafting. Gregg Division, McGraw-Hill Book Company, New York.

Wagener, J.L. 1980. FORTRAN 77 Principles of Programming. John Wiley & Sons, New York.

Walton, W.C. 1962. Selected Analytical Methods for Well and Aquifer Evaluation. Illinois State Water Survey. Bulletin 49.

Walton, W.C. 1985. Practical Aspects of Groundwater Modeling. Second Edition. National Water Well Association, Dublin, OH.

Wolverton, Van. 1984. Running MS DOS. Microsoft Press, Redmond, WA.

Index